Toroidal Dehn Fillings on Hyperbolic 3-Manifolds

Memoirs
of the
American Mathematical Society

Number 909

Toroidal Dehn Fillings on
Hyperbolic 3-Manifolds

Cameron McA. Gordon
Ying-Qing Wu

July 2008 • Volume 194 • Number 909 (end of volume) • ISSN 0065-9266

American Mathematical Society
Providence, Rhode Island

2000 *Mathematics Subject Classification.* Primary 57N10.

Library of Congress Cataloging-in-Publication Data

Gordon, Cameron, 1945–
  Toroidal Dehn fillings on hyperbolic 3-manifolds / Cameron McA. Gordon, Ying-Qing Wu.
    p. cm. — (Memoirs of the American Mathematical Society, ISSN 0065-9266 ; v, 194, no. 909)
  "Volume 194, number 909 (end of volume)."
  "July 2008."
  Includes bibliographical references.
  ISBN 978-0-8218-4167-9 (alk. paper)
   1. Dehn surgery (Topology) 2. Three-manifolds (Topology) 3. Geometry, Hyperbolic. I. Wu, Ying-Qing, 1956– II. Title.

QA649.G67 2008
514′.3—dc22   2008008511

Memoirs of the American Mathematical Society

This journal is devoted entirely to research in pure and applied mathematics.

Subscription information. The 2008 subscription begins with volume 191 and consists of six mailings, each containing one or more numbers. Subscription prices for 2008 are US$675 list, US$540 institutional member. A late charge of 10% of the subscription price will be imposed on orders received from nonmembers after January 1 of the subscription year. Subscribers outside the United States and India must pay a postage surcharge of US$38; subscribers in India must pay a postage surcharge of US$43. Expedited delivery to destinations in North America US$53; elsewhere US$130. Each number may be ordered separately; *please specify number* when ordering an individual number. For prices and titles of recently released numbers, see the New Publications sections of the *Notices of the American Mathematical Society*.

Back number information. For back issues see the *AMS Catalog of Publications*.

Subscriptions and orders should be addressed to the American Mathematical Society, P. O. Box 845904, Boston, MA 02284-5904, USA. *All orders must be accompanied by payment.* Other correspondence should be addressed to 201 Charles Street, Providence, RI 02904-2294, USA.

Copying and reprinting. Individual readers of this publication, and nonprofit libraries acting for them, are permitted to make fair use of the material, such as to copy a chapter for use in teaching or research. Permission is granted to quote brief passages from this publication in reviews, provided the customary acknowledgment of the source is given.

Republication, systematic copying, or multiple reproduction of any material in this publication is permitted only under license from the American Mathematical Society. Requests for such permission should be addressed to the Acquisitions Department, American Mathematical Society, 201 Charles Street, Providence, Rhode Island 02904-2294, USA. Requests can also be made by e-mail to reprint-permission@ams.org.

Memoirs of the American Mathematical Society (ISSN 0065-9266) is published bimonthly (each volume consisting usually of more than one number) by the American Mathematical Society at 201 Charles Street, Providence, RI 02904-2294, USA. Periodicals postage paid at Providence, RI. Postmaster: Send address changes to Memoirs, American Mathematical Society, 201 Charles Street, Providence, RI 02904-2294, USA.

© 2008 by the American Mathematical Society. All rights reserved.
Copyright of this publication reverts to the public domain 28 years
after publication. Contact the AMS for copyright status.
This publication is indexed in *Science Citation Index*®, *SciSearch*®, *Research Alert*®, *CompuMath Citation Index*®, *Current Contents*®/*Physical, Chemical & Earth Sciences*.
Printed in the United States of America.

∞ The paper used in this book is acid-free and falls within the guidelines
established to ensure permanence and durability.
Visit the AMS home page at http://www.ams.org/

10 9 8 7 6 5 4 3 2 1   13 12 11 10 09 08

Contents

1. Introduction — 1
2. Preliminary lemmas — 5
3. $\hat{\Gamma}_a^+$ has no interior vertex — 18
4. Possible components of $\hat{\Gamma}_a^+$ — 20
5. The case $n_1, n_2 > 4$ — 26
6. Kleinian graphs — 34
7. If $n_a = 4$, $n_b \geq 4$ and $\hat{\Gamma}_a^+$ has a small component then Γ_a is kleinian. — 37
8. If $n_a = 4$, $n_b \geq 4$ and Γ_b is non-positive then $\hat{\Gamma}_a^+$ has no small component — 41
9. If Γ_b is non-positive and $n_a = 4$ then $n_b \leq 4$ — 46
10. The case $n_1 = n_2 = 4$ and Γ_1, Γ_2 non-positive — 51
11. The case $n_a = 4$, and Γ_b positive — 54
12. The case $n_a = 2$, $n_b \geq 3$, and Γ_b positive — 64
13. The case $n_a = 2$, $n_b > 4$, Γ_1, Γ_2 non-positive, and $\max(w_1 + w_2, w_3 + w_4) = 2n_b - 2$ — 74
14. The case $n_a = 2$, $n_b > 4$, Γ_1, Γ_2 non-positive, and $w_1 = w_2 = n_b$ — 78
15. Γ_a with $n_a \leq 2$ — 85
16. The case $n_a = 2$, $n_b = 3$ or 4, and Γ_1, Γ_2 non-positive — 86
17. Equidistance classes — 94
18. The case $n_b = 1$ and $n_a = 2$ — 96
19. The case $n_1 = n_2 = 2$ and Γ_b positive — 97
20. The case $n_1 = n_2 = 2$ and both Γ_1, Γ_2 non-positive — 103
21. The main theorems — 108
22. The construction of M_i as a double branched cover — 111
23. The manifolds M_i are hyperbolic — 122
24. Toroidal surgery on knots in S^3 — 131

Bibliography — 139

Abstract

We determine all hyperbolic 3-manifolds M admitting two toroidal Dehn fillings at distance 4 or 5. We show that if M is a hyperbolic 3-manifold with a torus boundary component T_0, and r, s are two slopes on T_0 with $\Delta(r, s) = 4$ or 5 such that $M(r)$ and $M(s)$ both contain an essential torus, then M is either one of 14 specific manifolds M_i, or obtained from M_1, M_2, M_3 or M_{14} by attaching a solid torus to $\partial M_i - T_0$. All the manifolds M_i are hyperbolic, and we show that only the first three can be embedded into S^3. As a consequence, this leads to a complete classification of all hyperbolic knots in S^3 admitting two toroidal surgeries with distance at least 4.

Received by the editor December 1, 2005, and in revised form February 24, 2006.
2000 *Mathematics Subject Classification.* Primary 57N10.
Key words and phrases. Toroidal manifolds, Dehn fillings.
Cameron McA. Gordon was partially supported by NSF grant DMS 0305846.
Ying-Qing Wu was partially supported by NSF grant DMS 0203394.

1. Introduction

Let M be a *hyperbolic* 3-manifold, by which we shall mean a compact, connected, orientable 3-manifold such that M with its boundary tori removed admits a complete hyperbolic structure with totally geodesic boundary, and suppose that M has a torus boundary component T_0. If r is a slope on T_0 then $M(r)$ will denote the 3-manifold obtained by r-Dehn filling on M, i.e. attaching a solid torus V_r to M along T_0 in such a way that r bounds a disk in V_r. The Dehn filling $M(r)$ and the slope r are said to be *exceptional* if $M(r)$ is either reducible, ∂-reducible, annular, toroidal, or a small Seifert fiber space. Modulo the Geometrization Conjecture, the manifold $M(r)$ is hyperbolic if and only if $M(r)$ is not exceptional.

Thurston's Hyperbolic Dehn Surgery Theorem asserts that there are only finitely many exceptional Dehn fillings on each torus boundary component of M. It is known that if r, s are both exceptional then the geometric intersection number $\Delta = \Delta(r, s)$, also known as the *distance* between r and s, is small. In fact, the least upper bounds for Δ have been determined for all cases where neither $M(r)$ nor $M(s)$ is a small Seifert fiber space, by the work of many people. See [GW2] and the references therein.

For toroidal fillings, it was shown by Gordon [Go] that if r, s are toroidal slopes then $\Delta(r, s) \leq 8$, and moreover there are exactly two manifolds M with $\Delta = 8$, one with $\Delta = 7$, and one with $\Delta = 6$. In this paper we classify all the hyperbolic 3-manifolds which admit two toroidal Dehn fillings with $\Delta = 4$ or 5.

Already when $\Delta = 5$ there are infinitely many such manifolds. To see this, let M be the exterior of the Whitehead sister link, also known as the $(-2, 3, 8)$-pretzel link. The boundary of M consists of two tori T_0 and T_1, and there are slopes r, s on T_0 with $\Delta(r, s) = 5$ such that the Dehn filled manifolds $M(r) = M(r, *)$, $M(s) = M(s, *)$ are toroidal; see for example [GW3]. Now for infinitely many slopes t on T_1, $M_t = M(*, t)$ will be hyperbolic and $M_t(r) = M(r)(t)$, $M_t(s) = M(s)(t)$ will be toroidal. In this way we get infinitely many hyperbolic 3-manifolds with boundary a single torus having two toroidal fillings at distance 5. We shall show that, modulo this phenomenon, there are only finitely many M with two toroidal fillings at distance 4 or 5, and explicitly identify them. Define two triples (N_1, r_1, s_1) and (N_2, r_2, s_2) to be *equivalent*, denoted by $(N_1, r_1, s_1) \cong (N_2, r_2, s_2)$, if there is a homeomorphism from N_1 to N_2 which sends the boundary slopes (r_1, s_1) to (r_2, s_2) or (s_2, r_2).

THEOREM 1.1. *There exist 14 3-manifolds M_i, $1 \leq i \leq 14$, such that*

(1) M_i is hyperbolic, $1 \leq i \leq 14$;

(2) ∂M_i consists of two tori T_0, T_1 if $i \in \{1, 2, 3, 14\}$, and a single torus T_0 otherwise;

(3) there are slopes r_i, s_i on the boundary component T_0 of M_i such that $M(r_i)$ and $M(s_i)$ are toroidal, where $\Delta(r_i, s_i) = 4$ if $i \in \{1, 2, 4, 6, 9, 13, 14\}$, and $\Delta(r_i, s_i) = 5$ if $i \in \{3, 5, 7, 8, 10, 11, 12\}$;

(4) if M is a hyperbolic 3-manifold with toroidal Dehn fillings $M(r), M(s)$ where $\Delta(r, s) = 4$ or 5, then (M, r, s) is equivalent either to (M_i, r_i, s_i) for some $1 \leq i \leq 14$, or to $(M_i(t), r_i, s_i)$ where $i \in \{1, 2, 3, 14\}$ and t is a slope on the boundary component T_1 of M_i.

PROOF. The manifolds M_i are defined in Definition 21.3. (1) is Theorem 23.14. (2) follows from the definition. (3) and (4) follow from Theorem 21.4. □

REMARK 1.2. *Part (4) in Theorem 1.1 is still true if the hyperbolicity is replaced by the assumption that M is compact, connected, orientable, irreducible, atoroidal, and non Seifert fibered, in other words, M may be annular or ∂-reducible but not Seifert fibered.*

PROOF. First assume M is ∂-irreducible. Then any essential annulus must have at least one boundary component on a non-toroidal boundary component of M as otherwise M would be either toroidal or Seifert fibered. Attaching a hyperbolic manifold X to each non-toroidal boundary component of M will produce a hyperbolic manifold M', and we may choose X so that M' has more than three boundary components. One can show that $M'(r)$ and $M'(s)$ are still toroidal, which is a contradiction to Theorem 21.4. Now assume M is ∂-reducible. Then M is obtained by attaching 1-handles to the boundary of a manifold M''. If a 1-handle is attached to a toroidal component T of $\partial M''$ then either $M'' = T \times I$, which is impossible because $M(r)$ and $M(s)$ would be handlebodies and hence atoroidal, or T would be an essential torus in M, contradicting the assumption. It follows that M'' has a higher genus boundary component. One can check that $M''(r)$ and $M''(s)$ are still toroidal, which leads to a contradiction to Theorem 21.4 as above. □

The manifolds M_1, M_2 and M_3 were discussed in [GW1]; M_i, $i = 1, 2, 3$ is the exterior of a link L_i in S^3, where L_1 is the Whitehead link, L_2 is the 2-bridge link associated to the rational number $3/10$, and L_3 is the Whitehead sister link. See Figure 24.1. The other M_i can be built using intersection graphs on tori, see Definition 21.3 for more details. For $i \neq 4, 5$, each M_i can also be described as a double branched cover of a tangle $Q_i = (W_i, K_i)$, where W_i is a 3-ball for $i = 6, ..., 13$, and a once punctured 3-ball for $i = 1, 2, 3, 14$. This is done in [GW1] for $i = 1, 2, 3$, and in Section 22 for the other cases. See Lemma 22.2.

Some results on the case $\Delta = 5$ have been independently obtained by Teragaito [T2]. He obtains a finite set of pairs of intersection graphs of punctured tori at distance 5 which must contain all the pairs of graphs that arise from two toroidal fillings on a hyperbolic 3-manifold at distance 5. Seven of his pairs correspond to the manifolds in our list for $\Delta = 5$. The one in [T2, Figure 15] gives rise to a manifold obtained by attaching a thickened Möbius band to $T_1 \subset \partial M_3$ so that the union of the Möbius band with a Möbius band in $M_3(r_3)$ ($M_3(s_3)$) makes a Klein bottle. The manifold itself contains a $(2, 1)$ cable space and hence is non-hyperbolic, but one can attach a solid torus to it to make it hyperbolic. Therefore the graphs in [T2, Figure 15] correspond to infinitely many hyperbolic manifolds, all of type $M_3(t)$ for some slopes t on T_1.

We remark that the related problem of determining all hyperbolic 3-manifolds with two Dehn fillings at distance at least 4 that yield manifolds containing Klein bottles has been solved by Lee [L1, L2] (see also [MaS]).

Since M_i has more than one boundary component only when $i \in \{1, 2, 3, 14\}$, we have the following corollary to Theorem 1.1 (together with [Go]), which in the case $\Delta = 5$ is due to Lee [L1]. Note that all boundary components of the manifolds are tori.

COROLLARY 1.3. *Let M be a hyperbolic 3-manifold with more than one boundary component, having toroidal Dehn fillings $M(r), M(s)$ with $\Delta = \Delta(r, s) \geq 4$. Then each boundary component of M is a torus, and either*

(1) $\Delta = 4$ and $(M, r, s) \cong (M_i, r_i, s_i)$ for $i \in \{1, 2, 14\}$, or

(2) $\Delta = 5$ and $(M, r, s) \cong (M_3, r_3, s_3)$.

In [GW1] and [GW3] it is shown that if M is a hyperbolic 3-manifold with fillings $M(r)$ and $M(s)$, one of which is annular and the other either toroidal or annular, then either $(M, r, s) \cong (M_i, r_i, s_i)$ for $i \in \{1, 2, 3\}$, or $\Delta(r, s) \leq 3$. It is also known that if $M(r)$ contains an essential sphere or disk, and $M(s)$ contains an essential sphere, disk, annulus or torus, then $\Delta(r, s) \leq 3$; see [GW2] and the references listed there. Corollary 1.3 then gives

COROLLARY 1.4. *Let M be a hyperbolic 3-manifold with a torus boundary component T_0 and at least one other boundary component. Let r, s be exceptional slopes on T_0. Then either $(M, r, s) \cong (M_i, r_i, s_i)$ for $i \in \{1, 2, 3, 14\}$, or $\Delta(r, s) \leq 3$.*

In [GT] Goda and Teragaito showed that if a hyperbolic manifold M has at least three boundary components and has two toroidal slopes r, s then $\Delta(r, s) \leq 4$. Since each of the manifolds M_i in Theorem 1.1 has at most two boundary components, the above corollary shows that $\Delta(r, s) \leq 3$. This is sharp as there is an example in [GW1] that realizes this upper bound.

A pair (M, T_0) is called a *large manifold* if T_0 is a torus on the boundary of the 3-manifold M and $H_2(M, \partial M - T_0) \neq 0$ (see [Wu3]). Teragaito [T2] proved that there is no large hyperbolic manifold M admitting two toroidal fillings of distance at least 5. The following corollary clarifies the case of distance 4.

Theorem 22.3 *Suppose (M, T_0) is a large manifold and M is hyperbolic and contains two toroidal slopes r_1, r_2 on T_0 with $\Delta(r_1, r_2) \geq 4$. Then M is the Whitehead link exterior, and $\Delta(r_1, r_2) = 4$.*

Theorem 1.1 gives information about toroidal Dehn surgeries on hyperbolic knots in S^3. It follows from [Go] that the only such knot with two toroidal surgeries at distance > 5 is the figure eight knot, for which the 4 and -4 surgeries are toroidal. Teragaito has shown [T3] that the only hyperbolic knots with two toroidal surgeries at distance 5 are the Eudave-Muñoz knots $k(2, -1, n, 0)$, $n \neq 1$. We can now determine the knots with toroidal surgeries at distance 4. Denote by $L_i = K'_i \cup K''_i$ the link in Figure 24.1(i), where K'_i is the component on the left. Denote by $L_i(n)$ the knot obtained from K''_i by $1/n$ surgery on K'_i. One can check that $L_3(n)$ is the same as the Eudave-Muñoz knot $k(3, 1, -n, 0)$ in [Eu, Figure 25], which is the mirror-image of $k(2, -1, 1+n, 0)$ [Eu, Proposition 1.4].

Theorem 24.4 *A knot K in S^3 is hyperbolic and admits two toroidal surgeries $K(r_1), K(r_2)$ with $\Delta(r_1, r_2) \geq 4$ if and only if (K, r_1, r_2) is equivalent to one of the following, where n is an integer.*
  (1) $K = L_1(n)$, $n \neq 0, 1$; $r_1 = 0$, $r_2 = 4$.
  (2) $K = L_2(n)$, $n \neq 0, \pm 1$; $r_1 = 2 - 9n$, $r_2 = -2 - 9n$.
  (3) $K = L_3(n)$, $n \neq 0$; $r_1 = -9 - 25n$, $r_2 = -(13/2) - 25n$.
  (4) K is the Figure 8 knot; $r_1 = 4$, $r_2 = -4$.

The only hyperbolic knots known to have more than two toroidal surgeries are the figure eight knot and the $(-2, 3, 7)$-pretzel knot, with toroidal slopes $\{-4, 0, 4\}$ and $\{16, 37/2, 20\}$ respectively. This led Eudave-Muñoz [Eu] to conjecture that a hyperbolic knot in S^3 has at most three toroidal surgeries. Teragaito [T1] showed that there can be at most five toroidal surgeries. Theorem 1.1 and [T3, Corollary 1.2] lead to the following improvement.

Corollary 24.5 *A hyperbolic knot in S^3 has at most four toroidal surgeries. If there are four, then they are consecutive integers.*

Here is a sketch of the proof of Theorem 1.1. A toroidal Dehn filling $M(r)$ on a hyperbolic 3-manifold M gives rise to an essential punctured torus F in M whose boundary consists of $n > 0$ circles of slope r on T_0, where the capped-off surface $\hat{F}$ is an essential torus in $M(r)$. Hence, in the usual way (see Section 2), two toroidal fillings $M(r_1), M(r_2)$ give rise to a pair of intersection graphs Γ_1, Γ_2 on the tori $\hat{F}_1, \hat{F}_2$, with n_1, n_2 vertices respectively. The proof consists of a detailed analysis of the possible pairs of intersection graphs with $\Delta(r_1, r_2) = 4$ or 5, using Scharlemann cycles and other tools developed in earlier works in this area. This enables us to eliminate all but 17 pairs of graphs. As is usual in this kind of setting, the permissible graphs all have small numbers of vertices. Eleven of the pairs correspond to the manifolds M_i, $4 \leq i \leq 14$. We show that any of the remaining pairs must correspond to a pair of fillings on M_i or $M_i(t)$ for $i \in \{1, 2, 3\}$.

Here is a more detailed summary of the organization of the paper. Section 2 contains the basic definitions and some preliminary lemmas. In Sections 3-5 we deal with the generic case $n_1, n_2 > 4$, ultimately showing (Proposition 5.11) that this case cannot occur. More specifically, Section 3 shows that the reduced positive graph $\hat{\Gamma}_a^+$ of Γ_a (see Section 2 for definitions) has no interior vertices, and this is strengthened in Section 4 to showing that each component of $\hat{\Gamma}_a^+$ must be one of the 11 graphs in Figure 4.2. These are ruled out one by one in Section 5. In Sections 6-11 we consider the case where some $n_a = 4$. Section 6 discusses the situation where the graph Γ_a is *kleinian*; this arises when the torus $\hat{F}_a$ is the boundary of a regular neighborhood of a Klein bottle in $M(r_a)$. (The results here are also used in the discussion of the case $n_1, n_2 \leq 2$.) Sections 7, 8 and 9 show that if $n_a = 4$ and Γ_b is non-positive then $n_b \leq 4$. Section 10 shows that if Γ_1 and Γ_2 are both non-positive then $n_1 = n_2 = 4$ is impossible. Section 11 shows (Proposition 11.9) that if Γ_b is positive then there are exactly two pairs of graphs, one with $n_b = 2$, the other with $n_b = 1$. These give the manifolds M_4 and M_5 respectively. If we suppose $n_a \leq n_b$, it now easily follows (Proposition 11.10) that $n_a \leq 2$.

In Sections 12-16 we deal with the case $n_a \leq 2$, $n_b \geq 3$. The conclusion (Proposition 16.8) is that here there are exactly six pairs of graphs. Two of these are the ones described in Section 11, and the four new pairs give the manifolds M_6, M_7, M_8 and M_9. More precisely, in Section 12 we rule out the case where Γ_b is positive, and in Sections 13 and 14 we consider the case where $n_b > 4$ and both graphs Γ_1 and Γ_2 are non-positive. It turns out that here there is exactly one pair of graphs (Proposition 14.7), corresponding to the manifold M_6. We may now assume that $n_b = 3$ or 4. Section 15 establishes some notation and elementary properties for graphs with $n_a \leq 2$. In Section 16 we show that if Γ_1 and Γ_2 are non-positive then $n_b = 3$ is impossible and if $n_b = 4$ then there are exactly three examples, M_7, M_8 and M_9.

Sections 17-20 deal with the remaining cases where both n_1 and n_2 are ≤ 2. In Section 17 we introduce an equivalence relation, *equidistance*, on the set of edges of a graph Γ_a, and show that, under the natural bijection between the edges of Γ_1 and Γ_2, the two graphs induce the same equivalence relation. This gives a convenient way of ruling out certain pairs of graphs. Section 18 considers the case $n_a = 2$ and $n_b = 1$, and shows that here there are exactly three examples. Section 19 considers the case $n_1 = n_2 = 2$, Γ_b positive, showing that there are two examples. Finally, in Section 20 we consider the case $n_1 = n_2 = 2$, Γ_1 and Γ_2 both non-positive, and show that there are exactly six pairs of graphs in this case. The final list of all 11 possible pairs of graphs with $n_1, n_2 \leq 2$ is given in Proposition 20.4. Five of these correspond to the manifolds $M_{10}, M_{11}, M_{12}, M_{13}$ and M_{14}.

The remaining six pairs of graphs in Proposition 20.4 have the property that one of the graphs has a non-disk face. In Section 21 we show (Lemma 21.2), using the classification of toroidal/annular and annular/annular fillings at distance ≥ 4 given in [GW1] and [GW3], that in this case the manifold M is either M_1, M_2 or M_3, or is obtained from one of those by Dehn filling along one of the boundary components.

In Section 22 we show how the manifolds M_i, $6 \leq i \leq 14$ may be realized as double branched covers. Using this, in Section 23 we show that the manifolds M_i are hyperbolic. Finally, in Section 24 we give the applications to toroidal surgeries on knots in S^3.

We would like to thank Masakazu Teragaito and the referee for some very helpful comments.

2. Preliminary lemmas

Throughout this paper, we will fix a hyperbolic 3-manifold M, with a torus T_0 as a boundary component. A compact surface properly embedded in M is *essential* if it is π_1-injective, and is not boundary parallel. We use a, b to denote the numbers 1 or 2, with the convention that if they both appear in a statement then $\{a, b\} = \{1, 2\}$.

A slope on T_0 is a *toroidal slope* if $M(r_a)$ is toroidal. Let r_a be a toroidal slope on T_0. Denote by $\Delta = \Delta(r_1, r_2)$ the minimal geometric intersection number between r_1 and r_2. When $\Delta > 5$ the manifolds M have been determined in [Go]. We will always assume that $\Delta = 4$ or 5. Let $\hat{F}_a$ be an essential torus in $M(r_a)$, and let $F_a = \hat{F}_a \cap M$. If $M(r_a)$ is reducible then by [Wu1] and [Oh] we would have $\Delta \leq 3$, which is a contradiction. Therefore both $M(r_a)$ are irreducible.

Let n_a be the number of boundary components of F_a on T_0. Choose $\hat{F}_a$ in $M(r_a)$ so that n_a is minimal among all essential tori in $M(r_a)$. Minimizing the number of components of $F_1 \cap F_2$ by an isotopy, we may assume that $F_1 \cap F_2$ consists of arcs and circles which are essential on both F_a. Denote by J_a the attached solid torus in $M(r_a)$, and by u_i ($i = 1, ..., n_a$) the components of $\hat{F}_a \cap J_a$, which are all disks, labeled successively when traveling along J_a. Similarly let v_j be the disk components of $\hat{F}_b \cap J_b$. Let Γ_a be the graph on $\hat{F}_a$ with the u_i's as (fat) vertices, and the arc components of $F_1 \cap F_2$ as edges. Similarly for Γ_b. The minimality of the number of components in $F_1 \cap F_2$ and the minimality of n_a imply that Γ_a has no trivial loops, and that each disk face of Γ_a in $\hat{F}_a$ has interior disjoint from F_b.

If e is an edge of Γ_a with an endpoint x on a fat vertex u_i, then x is labeled j if x is in $u_i \cap v_j$. In this case e is called a *j-edge* in Γ_a, and an *i-edge* in Γ_b. Labels in Γ_a are considered as integers mod n_b; in particular, $n_b + 1 = 1$. When going around ∂u_i, the labels of the endpoints of edges appear as $1, 2, \ldots, n_b$ repeated Δ times. Label the endpoints of edges in Γ_b similarly.

Each vertex of Γ_a is given a sign according to whether J_a passes $\hat{F}_a$ from the positive side or negative side at this vertex. This also induces an orientation on the boundary of the vertex. Two vertices of Γ_a are *parallel* if they have the same sign, otherwise they are *antiparallel*. Note that if $\hat{F}_a$ is a separating surface, then n_a is even, and v_i, v_j are parallel if and only if i, j have the same parity. We use $val(v, G)$ to denote the valence of a vertex v in a graph G. If G is clear from the context, we simply denote it by $val(v)$.

When considering each family of parallel edges of Γ_a as a single edge $\hat{e}$, we get the *reduced graph* $\hat{\Gamma}_a$ on $\hat{F}_a$. It has the same vertices as Γ_a. Each edge of $\hat{\Gamma}_a$ represents a family of parallel edges in Γ_a. We shall often refer to a family of parallel edges as simply a *family*.

DEFINITION 2.1. *(1) An edge of Γ_a is a positive edge if it connects parallel vertices. Otherwise it is a negative edge.*

(2) The graph Γ_a is positive if all its vertices are parallel, otherwise it is non-positive.

We use Γ_a^+ (resp. Γ_a^-) to denote the subgraph of Γ_a whose vertices are the vertices of Γ_a and whose edges are the positive (resp. negative) edges of Γ_a. Similarly for $\hat{\Gamma}_a^+$ and $\hat{\Gamma}_a^-$.

A cycle in Γ_a consisting of positive edges is a *Scharlemann cycle* if it bounds a disk with interior disjoint from the graph, and all the edges in the cycle have the same pair of labels $\{i, i+1\}$ at their two endpoints, called the *label pair* of the Scharlemann cycle. A Scharlemann cycle containing only two edges is called a *Scharlemann bigon*. A Scharlemann cycle with label pair, say, $\{1, 2\}$ will also be called a (12)-Scharlemann cycle. If Γ_b contains a Scharlemann cycle with label pair $\{i, i \pm 1\}$, we shall sometimes abuse terminology and say that the vertex u_i of Γ_a is a *label of a Scharlemann cycle*. An *extended Scharlemann cycle* is a cycle of edges $\{e_1, ..., e_k\}$ such that there is a Scharlemann cycle $\{e'_1, ..., e'_k\}$ with e_i parallel and adjacent to e'_i and $e_i \neq e'_j$, $1 \leq i, j \leq k$. If $\{e_1, ..., e_k\}$ is a Scharlemann cycle in Γ_a then the subgraph of $\hat{\Gamma}_b$ consisting of these edges and their vertices is called a *Scharlemann cocycle*.

A subgraph G of a graph Γ on a surface F is *essential* if it is not contained in a disk in F. The following lemma contains some common properties of the graphs Γ_a. It can be found in [GW1, Lemma 2.2].

LEMMA 2.2. *(1) (The Parity Rule) An edge e is a positive edge in Γ_1 if and only if it is a negative edge in Γ_2.*

(2) A pair of edges cannot be parallel on both Γ_1 and Γ_2.

(3) If Γ_a has a set of n_b parallel negative edges, then on Γ_b they form mutually disjoint essential cycles of equal length.

(4) If Γ_a has a Scharlemann cycle, then $\hat{F}_b$ is separating. In particular, Γ_b has the same number of positive and negative vertices, so n_b is even, and two vertices v_i, v_j of Γ_b are parallel if and only if i, j have the same parity.

(5) If Γ_a has a Scharlemann cycle $\{e_1, \ldots, e_k\}$, then the corresponding Scharlemann cocycle on Γ_b is essential.

(6) If $n_b > 2$, then Γ_a contains no extended Scharlemann cycle.

Let $\hat{e}$ be a collection of parallel negative edges on Γ_b, oriented from v_1 to v_2. Then $\hat{e}$ defines a permutation $\varphi : \{1, \ldots, n_a\} \to \{1, \ldots, n_a\}$, such that an edge e in $\hat{e}$ has label k at v_1 if and only if it has label $\varphi(k)$ at v_2. Call φ the *transition function associated to* $\hat{e}$. Define the *transition number* to be the mod n_a integer $s = s(\hat{e})$ such that $\varphi(k) = k+s$. If we reverse the orientation of $\hat{e}$ then the transition function is φ^{-1}, and the transition number is $-s$; hence if $\hat{e}$ is unoriented then φ is well defined up to inversion, and $s(\hat{e})$ is well defined up to sign.

LEMMA 2.3. *(1) If a family of parallel negative edges in* Γ_a *contains more than* n_b *edges (in particular, if the family contains 3 edge endpoints with the same label), then* Γ_b *is positive, and the transition function associated to this family is transitive.*

(2) If Γ_a *contains two Scharlemann cycles with disjoint label pairs* $\{i, i+1\}$ *and* $\{j, j+1\}$, *then* $i \equiv j \mod 2$.

(3) If $n_b > 2$ *then a family of parallel positive edges in* Γ_a *contains at most* $n_b/2 + 2$ *edges, and if it does contain* $n_b/2 + 2$ *edges, then* $n_b \equiv 0 \mod 4$.

(4) Γ_a *has at most four labels of Scharlemann cycles, at most two for each sign.*

(5) A loop edge e *and a non-loop edge* e' *on* Γ_a *cannot be parallel on* Γ_b.

(6) If $n_b \geq 4$ *then* Γ_a *contains at most* $2n_b$ *parallel negative edges.*

PROOF. (1) This is obvious if $n_b \leq 2$, and it can be found in [GW1, Lemma 2.3] if $n_b > 2$.

(2) and (3) are basically Lemmas 1.7 and 1.4 of [Wu1]. If Γ_a has $n_b/2 + 2$ parallel positive edges then the two outermost pairs form two Scharlemann bigons. One can then check the labels of these Scharlemann bigons and use (2) to show that $n_b \equiv 0 \mod 4$.

(4) If Γ_a has more than four labels of Scharlemann cycles, then either one can find two Scharlemann cycles with disjoint label pairs $\{i, i+1\}$ and $\{j, j+1\}$ such that $i - j \equiv 1 \mod 2$, which is a contradiction to (2), or one can find three Scharlemann cycles with mutually disjoint label pairs, in which case one can replace $\hat{F}_a$ by another essential torus to reduce n_a and get a contradiction. See [Wu1, Lemma 1.10].

If Γ_a has three positive labels of Scharlemann cycles u_{i_j} then it has negative labels of Scharlemann cycles $u_{i_j + \epsilon_j}$ for some $\epsilon_j = \pm 1$, which cannot all be the same, hence Γ_a has at least 5 labels of Scharlemann cycles, contradicting the above.

(5) Since e is positive on Γ_a, it is negative in Γ_b. If e has endpoints on u_i in Γ_a then on Γ_b its two endpoints are both labeled i, hence the corresponding transition number is 0, so any edge e' parallel to e on Γ_b must also have the same label at its two endpoints, which implies that e' is a loop on Γ_a.

(6) This is [Go, Corollary 5.5]. □

LEMMA 2.4. *If a label* i *appears twice among the endpoints of a family* $\hat{e}$ *of parallel positive edges in* Γ_a, *then* i *is a label of a Scharlemann bigon in* $\hat{e}$. *In particular, if* $\hat{e}$ *has more than* $n_b/2$ *edges, then it contains a Scharlemann bigon.*

PROOF. Since the edges are positive, by the parity rule i cannot appear at both endpoints of a single edge in this family. Let $e_1, e_2, \ldots, e_k$ be consecutive edges of $\hat{e}$ such that e_1 and e_k have i as a label. Now k must be even, otherwise the edge $e_{(k+1)/2}$ would have the same label at its two endpoints. If $k \geq 4$ and $n_b > 2$ one can see that these edges contain an extended Scharlemann cycle, which contradicts Lemma 2.2(6). Therefore $k = 2$ or $n_b = 2$, in which case e_1, e_2 form a Scharlemann bigon with i as a label.

If $\hat{e}$ has more than $n_b/2$ edges, then it has more than n_b endpoints, so some label must appear twice. □

LEMMA 2.5. $\hat{\Gamma}_a$ *contains at most* $3n_a$ *edges.*

PROOF. Let V, E, F be the number of vertices, edges and disk faces of $\hat{\Gamma}_a$. Then $V - E + F \geq 0$ (the inequality may be strict if there are some non-disk faces.) Each face of $\hat{\Gamma}_a$ has at least three edges, hence we have $3F \leq 2E$. Solving those two inequalities gives $E \leq 3V$. □

LEMMA 2.6. *If $n_b > 4$ then the vertices of Γ_a cannot all be parallel.*

PROOF. By Lemma 2.5 the reduced graph $\hat{\Gamma}_a$ has at most $3n_a$ edges. For any i, since v_i on Γ_b has valence at least $4n_a$, there are $4n_a$ i-edges on Γ_a, hence two of them must be parallel, so i is a label of a Scharlemann cycle. Since there are at most 4 such labels (Lemma 2.3(4)), we would have $n_b \leq 4$, contradicting the assumption. □

A vertex v of a graph is a *full vertex* if all edges incident to it are positive.

LEMMA 2.7. *Suppose $n_b > 4$. Then*
(1) a family of parallel negative edges in Γ_b contains at most n_a edges, hence any label i appears at most twice among the endpoints of such a family;
(2) two families of positive edges in Γ_a adjacent at a vertex contain at most $n_b + 2$ edges; and
(3) three families of positive edges in Γ_a adjacent at a vertex contain at most $2n_b$ edges, and if there are $2n_b$ then $n_b = 6$.

PROOF. (1) If a family of parallel negative edges on Γ_b contains more than n_a edges then by Lemma 2.3(1) all vertices of Γ_a are parallel, which contradicts Lemma 2.6.

If i appears three times among the endpoints of a family of parallel negative edges in Γ_b then this family would contain more than n_a edges, which is a contradiction.

(2) By Lemma 2.3(3) a family of parallel positive edges contains $r \leq n_b/2 + 2$ edges. If two adjacent families $\hat{e}_1, \hat{e}_2$ contain more than $n_b + 2$ edges, then one of them, say $\hat{e}_1$, has $n_b/2 + 2$ edges while the other one has either $n_b/2 + 1$ or $n_b/2 + 2$ edges. Now $\hat{e}_1$ contains two Scharlemann bigons, which must appear on the two sides of the family because there is no extended Scharlemann cycle. There is also at least one Scharlemann bigon in $\hat{e}_2$. Examining the labels of these Scharlemann bigons we can see that they contain at least 5 labels, which contradicts Lemma 2.3(4).

(3) Assume the three families contain $r \geq 2n_b$ edges. Then one of the families contains more than $n_b/2$ edges, so by Lemma 2.2(4) n_b is even. By (2) two adjacent families of parallel edges contain at most $n_b + 2$ edges, while by Lemma 2.3(3) the other family has at most $n_b/2 + 2$ edges, so we have $2n_b \leq r \leq (n_b + 2) + (n_b/2 + 2)$, which gives $n_b \leq 8$.

If $n_b = 8$ then the above inequalities force the three families to have $6, 4, 6$ edges, and we see that all 8 labels appear as labels of Scharlemann bigons, which contradicts Lemma 2.3(4). So we must have $n_b = 6$. By Lemma 2.3(3) we have $2n_b \leq r \leq 3(n_b/2 + 1) = 12 = 2n_b$. Hence $r = 2n_b$. □

LEMMA 2.8. *If a vertex u_i of Γ_a is incident to more than n_b negative edges, then Γ_b has a Scharlemann cycle.*

PROOF. In this case there are $n_b + 1$ positive i-edges in Γ_b, which cut the surface F_b into faces, at least one of which is a disk face in the sense that it is a topological disk whose interior contains no vertices of Γ_b. Hence the subgraph of Γ_b consisting of these edges is a x-edge cycle in the sense of Hayashi-Motegi [HM, page 4468]. By [HM, Proposition 5.1] a disk face of this x-edge cycle contains a disk face of a Scharlemann cycle. □

Consider a graph G on a closed surface F, and assume that G has no isolated vertex. If the vertices of G have been assigned $\pm$ signs (for example $\hat{\Gamma}_a^+$), let X be the union of G and all its faces σ such that all vertices on $\partial\sigma$ have the same sign, otherwise let X be the union of G and all its disk faces. A vertex v of G is an *interior vertex* if it lies in the interior of X. A vertex v of G is a *cut vertex* if a regular neighborhood of v in X with v removed is not connected. A vertex v of G is a *boundary vertex* if it is not an interior or cut vertex. Note that if $G = \hat{\Gamma}_a^+$ then an interior vertex is a full vertex. Alternatively, let $\delta(v)$ be the number of corners around v which lie in X. Then v is an interior vertex if $\delta(v) = val(v, G)$, a boundary vertex if $\delta(v) = val(v, G) - 1$, and a cut vertex if $\delta(v) \leq val(v, G) - 2$.

Given a graph G on a surface D, let $c_i(G)$ be the number of boundary vertices of G with valence i. Define

$$\varphi(G) = 6c_0(G) + 3c_1(G) + 2c_2(G) + c_3(G)$$
$$\psi(G) = c_0(G) + c_1(G) + c_2(G) + c_3(G).$$

Note that $\psi(G)$ is the number of boundary vertices of G with valence at most 3.

LEMMA 2.9. *Let G be a connected reduced graph in a disk D such that any interior vertex of G has valence at least 6. Then $\varphi(G) \geq 6$. Moreover, if G is not homeomorphic to an arc or a single point then $\psi(G) \geq 3$.*

PROOF. Let X be the union of G and all its disk faces. The result is obviously true if G is a tree. So we assume that G has some disk faces.

First assume that X has no cut vertex, so it is a disk, and $c_0(G) = c_1(G) = 0$. The double of G along ∂X is then a graph $\tilde{G}$ on the double of X, which is a sphere. Note that the valence of a vertex v of $\tilde{G}$ is either at least 6, or it is 2 or 4 when v is a boundary vertex of G with valence 2 or 3, respectively. Since each face has at least three edges, an Euler characteristic argument gives

$$2 = V - E + F \leq V - \frac{1}{3}E = \sum_i (1 - \frac{1}{6}val(v_i, \tilde{G})) \leq \frac{2}{3}c_2(G) + \frac{1}{3}c_3(G).$$

Therefore $\varphi(G) = 2c_2(G) + c_3(G) \geq 6$. Since $c_0(G) = c_1(G) = 0$, we also have $\psi(G) = c_2(G) + c_3(G) \geq \frac{1}{2}\varphi(G) \geq 3$.

Now assume that X has a cut vertex v. Since G is connected and contained in a disk, X is simply connected, so we can write $X = X_1 \cup X_2$, where X_i are subcomplexes of X such that $X_1 \cap X_2 = v$, and $G_i = G \cap X_i$ are nontrivial connected subgraphs of G. The valence of v in G_i is at least 1, so its contribution to $\varphi(G_i)$ is at most 3. Hence by induction we have

$$\varphi(G) \geq (\varphi(G_1) - 3) + (\varphi(G_2) - 3) \geq 6.$$

By assumption X is not homeomorphic to an arc, so at least one of the X_i, say X_1, is not homeomorphic to an arc, and the other one has at least 2 boundary vertices of valence at most 3, whether it is homeomorphic to an arc or not. Hence

$$\psi(G) \geq \psi(G_1) + \psi(G_2) - 2 \geq 3 + 2 - 2 = 3.$$

□

LEMMA 2.10. *Let G be a reduced graph on a torus T with no interior or isolated vertex. Let V and E be the number of vertices and edges of G, and let k be the number of boundary vertices of G.*

(1) $k \geq E - V$, and equality holds if and only if all disk face of G are triangles, all non-disk faces are annuli, and each cut vertex has exactly two corners on annular faces.

(2) G has at most $2V$ edges.

PROOF. (1) Let D be the number of disk faces of G. Then $0 = \chi(T) \leq V - E + D$, and equality holds if and only if all non-disk faces are annuli. Thus $D \geq E - V$. For each vertex u of G, let $\delta(u)$ be the number of corners of disk faces incident to u. Then $\sum_u val(u) = 2E$, and $\sum_u \delta(u) \geq 3D \geq 3(E - V)$. Since there is no isolated or interior vertex, we have $val(u) - \delta(u) \geq 1$, and equality holds if and only if u is a boundary vertex. Let p be the number of non-boundary vertices. Then

$$p \leq \sum_u (val(u) - \delta(u) - 1) \leq 2E - 3(E - V) - V = 2V - E.$$

It follows that the number of boundary vertices is $k = V - p \geq E - V$, and equality holds if and only if (i) $V - E + D = 0$, i.e. non-disk faces are annuli, (ii) $\sum_u \delta(u) = 3D$, so all disk faces are triangles, and (iii) $val(u) - \delta(u) - 1 = 1$ for any cut vertex, i.e. each cut vertex has exactly two corners not on disk faces.

(2) Since the number of boundary vertices is at most V, by (1) we have $V \geq k \geq E - V$, hence $E \leq 2V$. □

LEMMA 2.11. *Suppose all interior vertices of $\hat{\Gamma}_a^+$ have valence at least 6, and all boundary vertices of $\hat{\Gamma}_a^+$ have valence at least 4. Let G be a component of $\hat{\Gamma}_a^+$. Then either (i) G is topologically an essential circle on the torus $\hat{F}_a$, or (ii) G has no cut vertex, all interior vertices of G are of valence exactly 6, and all boundary vertices of G are of valence exactly 4.*

PROOF. Let X be the union of G and all its disk faces. If X is the whole torus then all vertices are interior vertices, and an easy Euler characteristic argument shows that all vertices must be of valence 6, so (ii) follows. Also, by Lemma 2.9 X is not in a disk in $\hat{F}_a$ as otherwise G would have a boundary vertex of valence at most 3. Therefore we may assume that X has the homotopy type of a circle.

First assume that X has a cut vertex v. Recall that X is homotopy equivalent to a circle, so if $X - v$ is not connected, then v cuts off a subcomplex W of X which lies in a disk in $\hat{F}_a$. By Lemma 2.9 the graph $G \cap W$ has at least two boundary vertices of valence at most 3, hence at least one such vertex v' other than v, which contradicts the assumption because v' is then a boundary vertex of $\hat{\Gamma}_a^+$ of valence at most 3. Therefore we may assume that $X - v$ is connected. Since X has the homotopy type of a circle, X cut at v is a simply connected planar complex W, and X is obtained by identifying exactly two points of W. Let G' be the corresponding graph on W. We may assume that X is not a circle as otherwise (i) is true. Thus W is not homeomorphic to an arc. Therefore by Lemma 2.9 we have $\psi(G') \geq 3$, hence G' has at least one boundary vertex v' of valence at most 3 which is not identified to v in G. By definition v' is a boundary vertex of $\hat{\Gamma}_a^+$ of valence at most 3, which is a contradiction. This completes the proof that X has no cut vertex.

We may now assume that X is an annulus, so all vertices of G are either interior vertices of valence at least 6 in the interior of X, or boundary vertices of valence at least 4 on ∂X. Consider the double G'' of G on the double of X along ∂X. Since each boundary vertex of G of valence k gives rise to a vertex of valence $2k-2$ in G'', we see that G'' is a reduced graph on a torus such that all of its vertices have valence at least 6. An Euler characteristic argument shows that all vertices of G'' must have valence exactly 6, hence (ii) follows. □

LEMMA 2.12. *If $M(r_a)$ contains a Klein bottle K, then*
(1) $T = \partial N(K)$ is an essential torus in $M(r_a)$; and
(2) K intersects the core K_a of the Dehn filling solid torus at no less than $n_a/2$ points.

PROOF. T bounds a twisted I-bundle over the Klein bottle $N(K)$ on one side. Since $M(r_a)$ is assumed irreducible, if T is compressible on the other side then $M(r_a)$ is a Seifert fiber space over a sphere with (at most) three singular fibers of indices $(2,2,p)$ for some p, and if T is boundary parallel then $M(r_a)$ is a twisted I-bundle over the Klein bottle. Either case contradicts the assumption that $M(r_a)$ is toroidal. Therefore T is an essential torus. If $|K \cap K_a| < n_a/2$ then T would intersect K_a in less than n_a points, contradicting the choice of n_a. □

LEMMA 2.13. *Suppose $n_a > 2$, and Γ_b has both a 12-Scharlemann bigon $e_1 \cup e_2$ and a 23-Scharlemann bigon $e_3 \cup e_4$. If $e_1 \cup e_2$ and $e_3 \cup e_4$ are isotopic on $\hat{F}_a$, then the disk face D they bound on $\hat{F}_a$ contains at least $(n_a/2) - 1$ vertices in its interior.*

PROOF. Let m be the number of vertices in the interior of D. Let D_1, D_2 be the disk faces of (12)- and (23)-Scharlemann bigons in Γ_b. Shrinking the Dehn filling solid torus of $M(r_a)$ to its core K_a, the union $D_1 \cup D_2 \cup D$ is a Klein bottle Q in $M(r_a)$. A regular neighborhood of Q intersects K_a at an arc from u_1 to u_2 then to u_3, and one arc for each vertex of Γ_a in the interior of D. Hence Q can be perturbed to intersect K_a at $1+m$ points. By Lemma 2.12(2) we have $m+1 \geq n_a/2$, hence the result follows. □

An edge e of Γ_a is a *co-loop* edge if it has the same label on its two endpoints, in other words, it is a loop on the other graph Γ_b. Given a codimension 1 manifold X in a manifold Y, use $Y|X$ to denote the manifold obtained by cutting Y along X.

LEMMA 2.14. *Let $\hat{e}$ be a family of negative edges in Γ_a. Let G be the subgraph of Γ_b consisting of the edges of $\hat{e}$ and their vertices.*
(1) Each cycle component of G is an essential loop on $\hat{F}_b$.
(2) (The 3-Cycle Lemma.) G cannot contain three disjoint cycles; in particular, Γ_a cannot have three parallel co-loop edges.
(3) (The 2-Cycle Lemma.) If Γ_b is positive then G cannot contain two disjoint cycles; in particular, Γ_a cannot have two parallel co-loop edges.

PROOF. (1) Assume to the contrary that some cycle component of G is inessential on $\hat{F}_b$. Let D be a disk bounded by an innermost cycle component of G, and let D' be the bigon disks on F_a between edges of $\hat{e}$. Let V_b be the Dehn filling solid torus in $M(r_b)$. Then a regular neighborhood W of $D \cup V_b \cup D'$ is a solid torus containing the core of V_b as a cable knot winding along the longitude at least twice. See the proof of [GLi, Proposition 1.3]. In this case $W \cap M$ is a cable space, which is a contradiction to the assumption that M is a hyperbolic manifold.

(2) Let $\hat{e} = e_1 \cup ... \cup e_k$, oriented consistently, with tails at u' and heads at u'' on Γ_a. Let s be the transition number of $\hat{e}$. We may assume that e_i has label i at

its tail, so it has label $i + s$ at its head. Let D_j be the bigon on $\hat{F}_a$ between e_j and e_{j+1}.

If $k > n_b$ then by Lemma 2.3(1) the transition function associated with $\hat{e}$ has only one orbit, hence we may assume $k \leq n_b$. On Γ_b these edges form disjoint cycles and chains. Assume there are at least three cycles. Then e_1, e_2, e_3 belong to three distinct cycles C_1, C_2, C_3. Thus for $i = 1, 2, 3$,

$$C_i = e_i \cup e_{i+s} \cup \ldots \cup e_{i+(p-1)s}$$

is an oriented cycle on $\hat{F}_b$ for some fixed p. By (1) these are essential loops on $\hat{F}_b$, so they are parallel as unoriented loops.

Each bigon D_{1+js} gives a parallelism between an edge of C_1 and an edge of C_2, hence when shrinking the Dehn filling solid torus V_b to its core knot K_b, the union $A_1 = \cup D_{1+js}$ is an annulus in $M(r_b)$ with $\partial A_1 = C_1 \cup C_2$. Similarly, $A_2 = \cup D_{2+js}$ is an annulus in $M(r_b)$ with $\partial A_2 = C_2 \cup C_3$. These A_i are essential in $M(r_b)|\hat{F}_b$, the manifold obtained from $M(r_b)$ by cutting along $\hat{F}_b$, otherwise K_b would be isotopic to a curve having fewer intersections with $\hat{F}_b$.

Let A'_1, A'_2, A'_3 be the annuli $\hat{F}_b|(C_1 \cup C_2 \cup C_3)$, with $\partial A'_i = C_i \cup C_{i+1}$ (subscripts mod 3.) Let m_i be the number of times that K_b intersects the interior of A'_i. Then

$$\sum m_i + 3p = n_b$$

The annulus A_i is said to be of type I if a regular neighborhood of ∂A_i lies on the same side of $\hat{F}_b$, otherwise it is of type II. Note that if $\hat{F}_b$ is separating then A_i must be of type I. There are several possibilities. In each case one can find an essential torus T' in $M(r_b)$ which has fewer intersections with K_b. This will contradict the choice of $\hat{F}_b$ and complete the proof of (1).

Case 1. *C_2 is anti-parallel to both C_1 and C_3.*

In this case each $T_i = A_i \cup A'_i$ is a Klein bottle for $i = 1, 2$, which can be perturbed to intersect K_b at $p + m_i$ points. Since $\sum m_i + 3p = n_b$, either T_1 or T_2 can be perturbed to intersect K_b at fewer than $n_b/2$ points, contradicting Lemma 2.12.

Case 2. *C_2 is anti-parallel to C_1, say, and parallel to the other cycle C_3.*

Let $T_1 = A_1 \cup A'_1$ and $T_2 = A_1 \cup A_2 \cup A'_3$. Then T_i are Klein bottles, and they can be perturbed to intersect K_b at $p + m_1$ and m_3 points, respectively. One of these contradicts Lemma 2.12.

Case 3. *C_2 is parallel to both C_1 and C_3.*

If one of the A_i, say A_1, is of type II, then $T_1 = A_1 \cup A'_1$ is a non-separating torus (because it can be perturbed to intersect $\hat{F}_b$ transversely at a single circle), and it intersects K_b at $p + m_i < n_b$ points. Since $M(r_b)$ is irreducible, T_1 is incompressible and hence essential, which contradicts the choice of $\hat{F}_b$.

If both A_i are of type I then one can show that $A_1 \cup A_2 \cup A'_3$ is an essential torus T which can be perturbed to intersect K_b in $m_3 + p < n_b$ points. The proof is standard: The torus $\hat{F}_b$ and the annuli A_1, A_2 cut $M(r_b)$ into a manifold whose boundary contains four tori $T_1 = A_1 \cup A'_1$, $T_2 = A_1 \cup A'_2 \cup A'_3$, $T_3 = A_2 \cup A'_2$, and $T_4 = A_2 \cup A'_1 \cup A'_3$. Each of these tori T_i can be perturbed to have fewer than n_b intersections with the knot K_b, and hence bounds a manifold W_i which is either a solid torus or a $T^2 \times I$ between T_i and a component of $\partial M(r_b)$. Moreover, if W_i is a solid torus then the annulus $T_i \cap \hat{F}_b$ is essential on ∂W_i in the sense that

2. PRELIMINARY LEMMAS

it is neither meridional nor longitudinal (otherwise $\hat{F}_b$ would be compressible or could be isotoped to have fewer intersections with K_b). Now we have $M(r_b) = (W_1 \cup W_4) \cup (W_2 \cup W_3) = W' \cup W''$, with $W' \cap W''$ a torus $T = A_1 \cup A_2 \cup A'_3$ which can be perturbed to intersect K_b at $m_3 + p < n_b$ points. Since $W' = W_1 \cup_{A'_1} W_4$ and A'_1 is essential in both W_1 and W_4, T is incompressible and not boundary parallel in W'; similarly for W''. It follows that T is a contradiction to the choice of $\hat{F}_b$.

(3) The proof of this part is much simpler. Let A_1, A'_1 be as above, and let A''_1 be the complement of A'_1 on $\hat{F}_b$. If C_1, C_2 are parallel then $A_1 \cup A'_1$ is a nonseparating torus in $M(r_b)$ which can be perturbed to intersect K_b less than n_b times, contradicting the choice of $\hat{F}_b$. If C_1, C_2 are anti-parallel then $A_1 \cup A'_1$ and $A_1 \cup A''_1$ are Klein bottles, which can be perturbed to intersect K_b at a total of $n_b - 2p$ points, where p is the number of vertices in C_i; hence one of those will intersect K_b less than $n_b/2$ times, which contradicts Lemma 2.12. □

When studying Dehn surgery via intersection graphs, we usually fix the surfaces F_1, F_2, and hence the graphs Γ_1, Γ_2 are also fixed. The following technique will allow us to modify the surfaces and hence the graphs in certain situations. Lemma 2.15 will be used in the proofs of Lemmas 12.16 and 19.6.

Consider two surfaces F_1, F_2 in a 3-manifold M with boundary slopes r_1, r_2 respectively and suppose they intersect minimally. Let Γ_1, Γ_2 be the intersection graphs on $\hat{F}_1, \hat{F}_2$, respectively. Let α be a proper arc on a disk face D of Γ_a with boundary on edges of Γ_a. Then one can replace two small arcs of Γ_a centered at $\partial \alpha$ by two parallel copies of α to obtain a new graph Γ'_a, called the graph obtained from Γ_a by *surgery along* α.

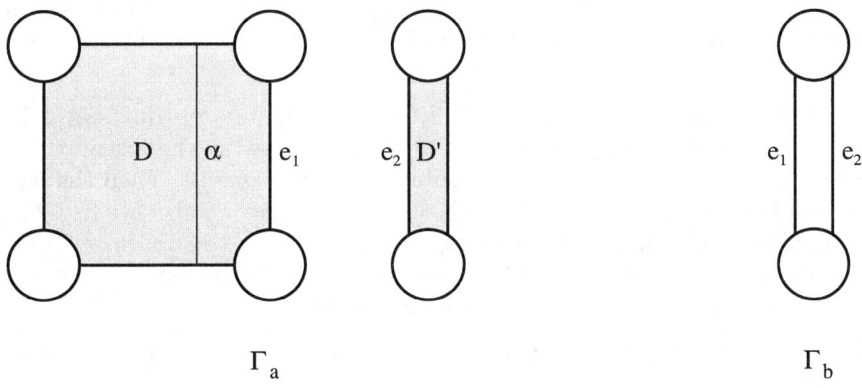

Figure 2.1

A face D' of Γ_a is called a *coupling face* to another face D of Γ_a along an edge e_1 of D if D' has an edge e_2 such that e_1, e_2 are adjacent parallel edges on Γ_b, and the neighborhoods in D and D' of the e_i's lie (locally) on the same side of $\hat{F}_b$. Note that this is independent of whether $\hat{F}_b$ is orientable or separating in M. See Figure 2.1. By definition D has no coupling face along e_1 if e_1 has no parallel edge on Γ_b, one coupling face along e_1 if e_1 has some parallel edges and is a border edge of the family, and two otherwise. A 4-gon face D of Γ_a looks like a "saddle surface" in $M|F_b$. In general it is not possible to push the saddle up or down to change the intersection graph. However, if some coupling face to an edge of D is a bigon then this is possible. See Figure 2.2. More explicitly, we have the following lemma.

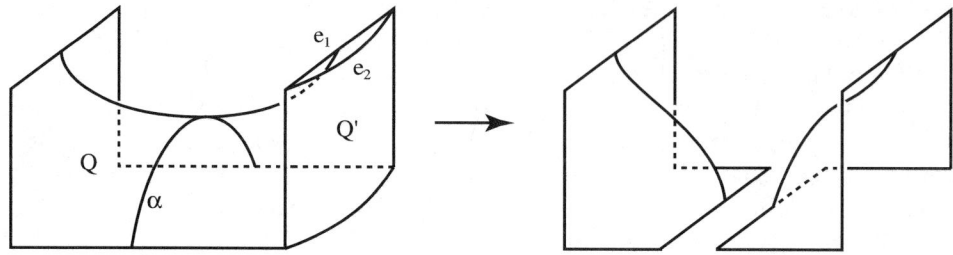

Figure 2.2

LEMMA 2.15. *Let Γ_1, Γ_2 be a pair of intersection graphs. Let Q be a face of Γ_a, and let e be an edge on ∂Q. Let α be an arc on Q with boundary in the interior of edges of Γ_a, cutting off a disk B_1 containing e and exactly two corners of Q. If some coupling face Q' of Q along e is a bigon, then F_a can be isotoped so that the new intersection graph Γ'_a is obtained from Γ_a by surgery along α.*

PROOF. Cut M along F_b. Then the face Q is as shown in Figure 2.2. Let B_2 be the bigon in Γ_b between e and the edge e' on Q'. After shrinking the Dehn filling solid torus V_b to its core knot K_b, the union $B_1 \cup B_2 \cup Q'$ is a disk Q'' with boundary the union of α and an arc on $\hat{F}_b$. Pushing Q'' off K_b gives a disk P in M which has boundary the union of α and an arc on F_b, and has interior disjoint from $F_b \cup F_a$. Therefore we can isotope F_a through this disk P to get a new surface F'_a. It is clear that the new intersection graph Γ'_a is obtained from Γ_a by surgery along α. □

Let u be a vertex of Γ_a, and P, Q two edge endpoints on ∂u. Let I be the interval on ∂u from P to Q along the direction induced by the orientation of u. The edge endpoints of Γ_a cut I into k subintervals for some k. Then the *distance* from P to Q on ∂u is defined as $d_u(P, Q) = k$. Sometimes we also use $d_{\Gamma_a}(P, Q)$ to denote $d_u(P, Q)$. If P, Q are the only edge endpoints of e_1, e_2 on ∂u, respectively, then we define $d_u(e_1, e_2) = d_u(P, Q)$. Notice that if the valence of u is m, then $d_u(Q, P) = m - d_u(P, Q)$. The following lemma can be found in [Go].

LEMMA 2.16. [Go, Lemma 2.4] *(i) Suppose $P, Q \in \partial u_i \cap \partial v_k$ and $R, S \in \partial u_j \cap \partial v_l$. If $d_{u_i}(P, Q) = d_{u_j}(R, S)$ then $d_{v_k}(P, Q) = d_{v_l}(R, S)$.*

(ii) Suppose that $P \in u_i \cap v_k$, $Q \in u_i \cap v_l$, $R \in u_j \cap v_k$, and $S \in u_j \cap v_l$. If $d_{u_i}(P, Q) = d_{u_j}(R, S)$, then $d_{v_k}(P, R) = d_{v_l}(Q, S)$.

Suppose two edges e_1, e_2 of Γ_a connect the same pair of vertices u_i, u_j. Let p_k, q_k be the endpoints of e_k on u_i, u_j, respectively, $k = 1, 2$. Then e_1, e_2 are *equidistant* if $d_{u_i}(p_1, p_2) = d_{u_j}(q_2, q_1)$. (Note that the orders of the edge endpoints have been reversed.) Thus for example a pair of parallel positive edges is always equidistant, but a pair of parallel negative edges is not unless their distance is exactly half of the valence of the vertices.

Note that when $u_i \neq u_j$ the above equation can be written as $d_{u_i}(e_1, e_2) = d_{u_j}(e_2, e_1)$. When $u_i = u_j$, $d_{u_i}(e_1, e_2)$ is not defined, and there are two choices for the pair p_k, q_k, but one can check that whether the equality $d_{u_i}(p_1, p_2) = d_{u_j}(q_2, q_1)$ holds is independent of the choice of p_i, q_i.

The following lemma is called the *Equidistance Lemma*. It follows from Lemma 2.16, and can also be found in [GW1]. Given an edge e, define ∂e to be the pair of vertices at the endpoints of e.

LEMMA 2.17. [GW1, Lemma 2.8] *Let e_1, e_2 be a pair of edges with $\partial e_1 = \partial e_2$ in both Γ_1 and Γ_2. Then e_1, e_2 are equidistant in Γ_1 if and only if they are equidistant in Γ_2.*

Given two oriented slopes r_1, r_2 on T_0, choose an oriented meridian-longitude pair m, l on the torus T_0 so that $r_1 = m$, then the slope r_2 is homologous to $Jm + \Delta l$ for some mod Δ integer $J = J(r_1, r_2)$, called the *jumping number* between r_1, r_2. Note that if $\Delta = 4$, then $J = \pm 1$, and if $\Delta = 5$, then $J = \pm 1$ or ± 2. The following lemma is call the *Jumping Lemma* and can be found in [GW1].

LEMMA 2.18. [GW1, Lemma 2.10] *Let $P_1, \ldots, P_\Delta$ be the points of $\partial u_i \cap \partial v_j$, labeled successively on ∂u_i. Let $J = J(r_1, r_2)$ be the jumping number of r_1, r_2. Then on v_j these points appear in the order $P_J, P_{2J}, \ldots, P_{\Delta J}$. In particular, they appear successively as $P_1, \ldots, P_\Delta$ along some direction of ∂v_j if and only if $J = \pm 1$.*

LEMMA 2.19. *Let $e_1 \cup \ldots \cup e_p$ and $e'_1 \cup \ldots \cup e'_q$ be two sets of parallel edges on Γ_a. Suppose e_1 is parallel to e'_1 and e_p parallel to e'_q on Γ_b. Then $p = q$.*

PROOF. Let D_1, D_2, D_3, D_4 be the disks realizing the parallelisms of $e_1 \cup e_p$ and $e'_1 \cup e'_q$ on Γ_a, and $e_1 \cup e'_1$ and $e_p \cup e'_q$ on Γ_b. Then the union $A = D_1 \cup \ldots \cup D_4$ is a Möbius band or annulus in M with boundary on T_0. (It is embedded in M, otherwise there is a pair of edges parallel in both graphs, contradicting Lemma 2.2(2).) If A is a Möbius band then it is already a contradiction to the hyperbolicity of M. If A is an annulus and $p \neq q$ then a boundary component c of A has intersection number $p - q \neq 0$ with $\cup \partial v_i$ and hence is an essential curve on T_0. Since e_1 is an essential arc on both A and F_a and F_a is boundary incompressible, A cannot be boundary parallel. It follows that A is an essential annulus in M, which again contradicts the assumption that M is hyperbolic. □

LEMMA 2.20. *Suppose Γ_b is positive, $n_b \geq 3$, and Γ_a contains bigons $e_1 \cup e_2$ and $e'_1 \cup e'_2$, such that e_1, e'_1 have label pair $\{i, j\}$ and e_2, e'_2 have label pair $\{i+1, j+1\}$, where $j \neq i$. Let $C_1 = e_1 \cup e'_1$ and $C_2 = e_2 \cup e'_2$ be the loops on $\hat{F}_b$. If C_1 is essential on $\hat{F}_b$ then C_2 is essential on $\hat{F}_b$ and not homotopic to C_1.*

PROOF. Let B and B' be the bigon faces bounded by $e_1 \cup e_2$ and $e'_1 \cup e'_2$, respectively. Shrinking the Dehn filling solid torus to the core knot K_b, the union $B \cup B'$ becomes an annulus A_1 in $M(r_b)$ with boundary $C_1 \cup C_2$. Since $\hat{F}_b$ is incompressible and C_1 is essential on $\hat{F}_b$, it follows that C_2 must also be essential on $\hat{F}_b$.

Now assume C_i are essential and homotopic on $\hat{F}_b$. Since $i \neq j$ and $n_b > 2$, C_1, C_2 have at most one vertex in common. If C_1, C_2 are disjoint, let A_2 be an annulus on $\hat{F}_b$ bounded by $C_1 \cup C_2$. If C_1, C_2 has a common vertex $v_{i+1} = v_j$, let A_2 be the disk face of $C_1 \cup C_2$ in $\hat{F}_b$, which will be considered as a degenerate annulus as it can be obtained from an annulus by pinching an essential arc to a point. Let A'_2 be the closure of $\hat{F}_b - A_2$. Let m and m' be the number of vertices in the interior of A_2 and A'_2, respectively. Then $n_b = m + m' + k$, where k is the number of vertices on $C_1 \cup C_2$, i.e., $k = 4$ if $C_1 \cap C_2 = \emptyset$, and $k = 3$ otherwise.

First consider the case that $C_1 \cap C_2 = \emptyset$. Orient C_1, C_2 so that they are parallel on the annulus A_1. If they are also parallel on $\hat{F}_b$ then $A_1 \cup A_2$ is a nonseparating torus which can be perturbed to intersect K_b at $m + 2 < n_b$ points, which is a contradiction. If they are anti-parallel then $A_1 \cup A_2$ and $A_1 \cup A'_2$ are Klein bottles which can be perturbed to intersect K_b at m and m' points, respectively. Since at least one of m, m' is less than $n_b/2$, this contradicts Lemma 2.12.

The case that $C_1 \cap C_2 \neq \emptyset$ is similar. If C_1, C_2 are parallel then $A_1 \cup A_2$ is a torus and can be perturbed to intersect K_b at $m + 1 < n_b$ points; if they are anti-parallel then $A_1 \cup A_2$ and $A_1 \cup A'_2$ can be perturbed to be Klein bottles intersecting K_b at $m + 1$ and m' points, respectively, which leads to contradictions as above because $m + 1 + m' < n_b$ implies either $2(m+1) < n_b$ or $2m' < n_b$. □

A triple of edge endpoints (p_1, p_2, p_3) on Γ_b is *positive* if they appear on the boundary of the same vertex v_i, and in this order on ∂v_i along the orientation of ∂v_i. Note that this is true if and only if $d_{v_i}(p_1, p_2) + d_{v_i}(p_2, p_3) = d_{v_i}(p_1, p_3)$.

LEMMA 2.21. *(1) Suppose (p_1, p_2, p_3) is a positive triple on Γ_b. Let k be a fixed integer and let p'_i be edge endpoints such that $d_{\Gamma_a}(p_i, p'_i) = k$ for all i. Then (p'_1, p'_2, p'_3) is also a positive triple on Γ_b.*

(2) Let $e_1 \cup ... \cup e_r$ be a set of parallel negative edges with end vertices u_1, u_2 in Γ_a. Let $u(p) \in \{u_1, u_2\}$ for $p = 1, 2, 3$, and let $e_j(u(p))$ be the endpoint of e_j at $u(p)$. If $(e_i(u(1)), e_j(u(2)), e_k(u(3)))$ is a positive triple and $i, j, k \leq r - t$, then $(e_{i+t}(u(1)), e_{j+t}(u(2)), e_{k+t}(u(3)))$ is also a positive triple.

PROOF. (1) Geometrically this is obvious: Flowing on T_0 along ∂F_a moves the first triple to the second triple, hence the orientations of the components of ∂F_b containing these triples are the same on T_0.

Alternatively one may use Lemma 2.16(ii) to prove the result. Since $d_{\Gamma_a}(p_i, p'_i) = d_{\Gamma_a}(p_j, p'_j) = k$ for all i, j, by Lemma 2.16(ii) we have $d_{\Gamma_b}(p_i, p_j) = d_{\Gamma_b}(p'_i, p'_j)$ for all i, j. Therefore $d_{\Gamma_b}(p_1, p_2) + d_{\Gamma_b}(p_2, p_3) = d_{\Gamma_b}(p_1, p_3)$ if and only if $d_{\Gamma_b}(p'_1, p'_2) + d_{\Gamma_b}(p'_2, p'_3) = d_{\Gamma_b}(p'_1, p'_3)$.

(2) This is a special case of (1) because

$$\begin{aligned} d_{\Gamma_a}(e_i(u(1)), e_{i+t}(u(1))) &= d_{\Gamma_a}(e_j(u(2)), e_{j+t}(u(2))) \\ &= d_{\Gamma_a}(e_k(u(3)), e_{k+t}(u(3))) = t \end{aligned}$$

□

LEMMA 2.22. *Suppose Γ_b is positive and $n = n_b \geq 3$.*

(1) Suppose $\hat{e} \supset e_1 \cup ... \cup e_{n+2}$, and the transition number $s = 1$. Let A be the annulus obtained by cutting $\hat{F}_b$ along the cycle $e_1 \cup ... \cup e_n$. Then the edges e_{n+1}, e_{n+2} lie in A as shown in Figure 2.3, up to reflection along the center circle of the annulus.

(2) If $s = 1$, $n = 3$ and $\hat{e}_1$ contains 6 edges $e_1 \cup ... \cup e_6$ then the edges are as shown in Figure 2.4.

(3) Any family of parallel negative edges in Γ_a contains at most $2n$ edges.

PROOF. (1) Let u, u' be the end vertices of $\hat{e}$ in Γ_a. Orient e_i from u to u' and assume without loss of generality that e_i has label i at its tail $e_i(t)$ in Γ_a. Since $s = 1$, the head of e_i, denoted by $e_i(h)$, has label $i + 1$ in Γ_a. The edges $e_1, ..., e_n$

form an essential loop on the torus $\hat{F}_b$. Cutting $\hat{F}_b$ along this loop produces an annulus, as shown in Figure 2.3.

Up to reflection along the center circle of the annulus we may assume that the edge e_{n+1} appears in this annulus as shown in Figure 2.3. We need to prove that e_{n+2} appears in Γ_b as shown in the figure.

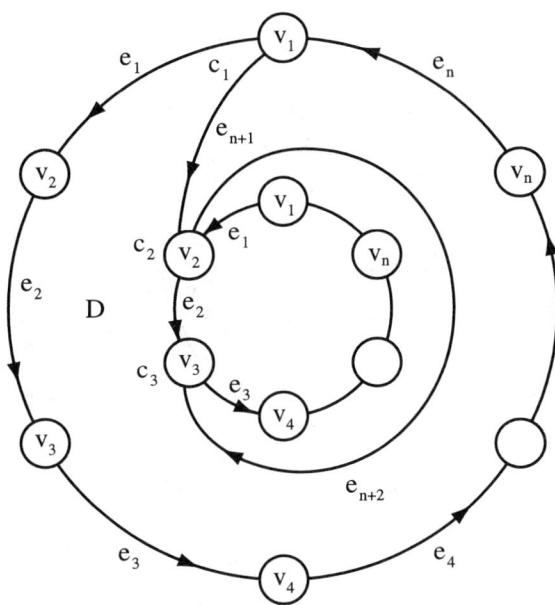

Figure 2.3

Since Γ_b is positive, we may assume that all vertices on Figure 2.3 are oriented counterclockwise. Note that $(e_1(h), e_{n+1}(h), e_2(t))$ is a positive triple on Γ_b. By Lemma 2.21(2) the triple $(e_2(h), e_{n+2}(h), e_3(t))$ is also a positive triple. This determines the location of the head of e_{n+2}, as shown in Figure 2.3. Applying Lemma 2.20 to $e_1 \cup e_2$ and $e_{n+1} \cup e_{n+2}$, we see that the loop $e_2 \cup e_{n+2}$ is essential and not homotopic to $e_1 \cup e_{n+1}$ on $\hat{F}_b$, so these two loops must intersect transversely at the common vertex v_2 on $\hat{F}_b$. Hence the edge e_{n+2} must appear as shown in Figure 2.3.

(2) By (1) the first 5 edges must be as shown in Figure 2.4. These cut the torus into a 3-gon and a 7-gon. The edge e_6 is not parallel to the other e_i's on Γ_b and hence must lie in the 7-gon, connecting v_3 to v_1. For the same reason as above, $(e_3(h), e_6(h), e_4(t))$ is a positive triple on ∂v_1, hence the head of e_6 must be in the corner on ∂v_1 from $e_1(t)$ to $e_4(t)$ because the corner from $e_3(h)$ to $e_1(t)$ lies in the 3-gon. Similarly, since $(e_1(h), e_5(t), e_4(h))$ is a positive triple on Figure 2.4, by Lemma 2.21(2) $(e_2(h), e_6(t), e_5(h))$ is also a positive triple, which determines the position of the tail of e_6. Therefore e_6 must be as shown in Figure 2.4.

(3) This follows from [Go, Corollary 5.5] when $n \geq 4$. Now assume $n = 3$ and suppose there exist $2n + 1 = 7$ parallel edges $e_1 \cup ... \cup e_7$ on Γ_a. By the 3-Cycle Lemma 2.14(2) we may assume that the transition number $s \neq 0$. Since $n = 3$, we may assume without loss of generality that $s = 1$, hence by (2) the subgraph of Γ_b consisting of the edges $e_1 \cup ... \cup e_6$ is as shown in Figure 2.4. By the same argument

as above, $(e_4(h), e_7(h), e_5(t))$ and $(e_3(h), e_7(t), e_6(h))$ are positive triples on ∂v_2 and ∂v_1, respectively. Since e_7 must lie in the 6-gon face D in Figure 2.4, this is possible only if e_7 is parallel to e_1, which is a contradiction to Lemma 2.2(2). □

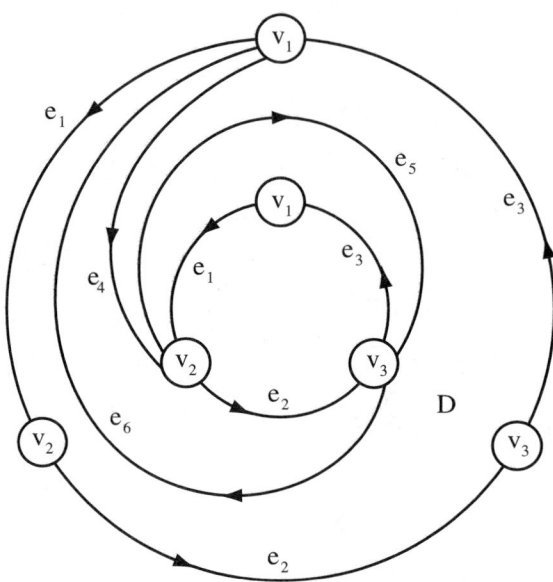

Figure 2.4

LEMMA 2.23. *Let sign(v) be the sign of a vertex v, and define p_a to be the sum of the signs of the vertices of Γ_a. Then either $p_1 = 0$ or $p_2 = 0$. In particular, n_1, n_2 cannot both be odd.*

PROOF. For each edge endpoint c on $u_i \cap v_j$, define $\text{sign}(c) = \text{sign}(u_i)\,\text{sign}(v_j)$. Then the parity rule says that the two endpoints of an edge e have different sign. Summing over all edge endpoints on Γ_a gives

$$0 = \sum_{i,j} \Delta(\text{sign}(u_i)\,\text{sign}(v_j)) = \Delta \sum_i \text{sign}(u_i) \sum_j \text{sign}(v_j) = \Delta p_1 p_2$$

hence either $p_1 = 0$ or $p_2 = 0$. □

3. $\hat{\Gamma}_a^+$ has no interior vertex

In this section we will show that if $n_b > 4$ then the graph Γ_a does not have interior vertices; in particular the vertices of Γ_a cannot all be parallel. Recall that we have assumed that $\Delta \geq 4$.

LEMMA 3.1. *If $n_b > 4$ then $\hat{\Gamma}_a$ has no full vertex of valence at most 6.*

PROOF. First assume $n_b \neq 6$. Then by Lemma 2.7(3), three adjacent families of positive edges in Γ_a contain at most $2n_b - 1$ edges, hence if Γ_a has a full vertex of valence at most 6 then

$$4n_b \leq \Delta n_b \leq 2(2n_b - 1) = 4n_b - 2,$$

a contradiction.

So suppose $n_b = 6$, and let u_1 be a full vertex of $\hat{\Gamma}_a$ of valence at most 6. By Lemma 2.3(3), each family of edges incident to u_1 contains at most 4 edges. Since there are at least 24 edges in at most 6 families, there must be exactly 6 families, each containing exactly 4 edges. Each family contains a Scharlemann bigon, so there are six Scharlemann bigons at the vertex u_1. Since there are no extended Scharlemann cycles, each Scharlemann bigon appears at one end of a family of parallel edges. Thus by examining the labels around the vertex u_1, one can see that if one Scharlemann bigon has label pair $\{1,2\}$ then the others must have label pair $\{1,2\}$, $\{3,4\}$ or $\{5,6\}$, and that at least two pairs do occur as label pairs of Scharlemann bigons. On the other hand, by Lemma 2.3(4) all three pairs cannot appear as label pairs of Scharlemann bigons. Hence, without loss of generality, there are incident to u_1 at least three (12)-Scharlemann bigons and a (34)-Scharlemann bigon. Since on Γ_b the edges of the (34)-Scharlemann bigon form an essential loop on $\hat{\Gamma}_b$, there are at most two edges of $\hat{\Gamma}_b$ joining v_1 to v_2. Since the three (12)-Scharlemann bigons give rise to six negative 1-edges of Γ_b joining v_1 to v_2, three of these must be parallel, contradicting Lemma 2.7(1). □

LEMMA 3.2. *If $n_b > 4$ then $\hat{\Gamma}_a^+$ has no interior vertices.*

PROOF. This follows from Lemma 3.1 if $n_a \leq 2$ because in this case either $\hat{\Gamma}_a^+ = \hat{\Gamma}_a$ and there is a full vertex of valence at most 6, or $n_a = 2$ and there is no interior vertex. Therefore we may assume that $n_a \geq 3$.

Suppose to the contrary that $\hat{\Gamma}_a^+$ has an interior vertex u_i. By Lemma 3.1 all interior vertices of $\hat{\Gamma}_a^+$ have valence at least 7, hence we can apply Lemma 2.11 to conclude that $\hat{\Gamma}_a^+$ has a boundary vertex u_1 of valence at most 3.

By Lemma 2.7(3) the three families of adjacent positive edges at u_1 contain at most $2n_b$ edges, hence there are $2n_b$ adjacent negative edges. On Γ_b this implies that each vertex v_j is incident to two positive edges with label 1 at v_j, which cannot be parallel as otherwise there would be at least $n_a + 1 > n_a/2 + 2$ parallel positive edges, contradicting Lemma 2.3(3). Therefore the reduced graph $\hat{\Gamma}_b$ contains at least n_b positive edges. On the other hand, the existence of an interior vertex in $\hat{\Gamma}_a^+$ implies that $\hat{\Gamma}_b$ contains at least $2n_b$ negative edges, as shown in the proof of Lemma 3.1. Since $\hat{\Gamma}_b$ has at most $3n_b$ edges (Lemma 2.5), it must have exactly n_b positive edges and $2n_b$ negative edges. Since we have shown above that each vertex in $\hat{\Gamma}_b$ is incident to at least two positive edges, it follows that it is incident to exactly two positive edges.

We claim that a family of parallel positive edges in Γ_b contains at most $n_a/2$ edges. If such a family contains more than $n_a/2$ edges, then there is a Scharlemann bigon on one side of the family, and by looking at the labels one can see that all labels appear among the endpoints of this family, which is impossible because u_i being an interior vertex in $\hat{\Gamma}_a^+$ implies that all edges in Γ_b with i as a label are negative.

Since each vertex v_j is incident to two families of positive edges, each containing at most $n_a/2$ edges, we see that v_j is incident to at least $3n_a$ negative edges. By Lemmas 3.1 and 2.5 we see that $\hat{\Gamma}_a$ has less than $3n_a$ positive edges, hence two of the negative edges incident to v_j are parallel in Γ_a, so j is a label of a Scharlemann bigon in Γ_a. Since this is true for all vertices in Γ_b, by Lemma 2.3(4) we have $n_b \leq 4$, which is a contradiction. □

4. Possible components of $\hat{\Gamma}_a^+$

LEMMA 4.1. *Suppose $\hat{\Gamma}_a^+$ has no isolated vertex or interior vertex. If some v_i of Γ_b is incident to more than $2n_a$ negative edges in Γ_b, or if $n_a > 4$ and v_i is incident to at most two families of positive edges in Γ_b, then i is a label of a Scharlemann bigon in Γ_a.*

PROOF. If v_i is a vertex of Γ_b incident to more than $2n_a$ negative edges then by Lemma 2.10(2) two of them are parallel in Γ_a, so by Lemma 2.4 they form a Scharlemann bigon, hence i is the label of a Scharlemann bigon in Γ_a. If $n_a > 4$ and v_i is incident to two families of positive edges in Γ_b then by Lemma 2.3(3) each family contains less than n_a edges, hence v_i is incident to more than $2n_a$ negative edges and the result follows from the above. □

In the rest of this section we assume $n_a > 4$ for $a = 1, 2$, and $\Delta \geq 4$. By Lemma 3.2 $\hat{\Gamma}_a^+$ has no interior vertices. We will show that each component of $\hat{\Gamma}_a^+$ must be one of the 11 graphs in Figure 4.2.

LEMMA 4.2. *No vertex u of $\hat{\Gamma}_a$ is incident to at most four positive edges and at most one negative edge.*

PROOF. By Lemmas 2.7(1) and 2.7(2) a family of negative edges contains at most n_b edges, and four adjacent families of positive edges contain at most $2(n_b + 2) = 2n_b + 4$ edges. Since Γ_a has at least $4n_b$ edges incident to u, we would have $n_b \leq 4$, which is a contradiction to our assumption. □

LEMMA 4.3. *Suppose u_i is incident to at most three positive edges in $\hat{\Gamma}_a$, and if there are three then two of them are adjacent. Then i is a label of a Scharlemann bigon in Γ_b.*

PROOF. In this case each label appears at the endpoint of some negative edge at u_i, so $\hat{\Gamma}_b^+$ has no isolated vertex. By Lemma 4.1 the result is true if u_i is incident to more than $2n_b$ negative edges. So we assume that u_i is incident to no more than $2n_b$ negative edges, and hence at least $2n_b$ positive edges. By Lemma 2.7(2) the two adjacent families of positive edges contain at most $n_b + 2$ edges, while the other positive family contains no more than $n_b/2 + 2$ edges. Thus $(n_b + 2) + (n_b/2 + 2) \geq 2n_b$, which gives $n_b \leq 8$. Since one of the positive families contains more than $n_b/2$ edges, it contains a Scharlemann bigon; by Lemma 2.2(4) n_b must be even, so $n_b = 8$ or 6. Using the above inequality and the fact that when $n_b = 6$ each positive family contains at most 4 edges (Lemma 2.3(3)), we see that u_i is incident to exactly $2n_b$ positive edges and $2n_b$ negative edges. Dually, this implies that in Γ_b there are exactly $2n_b$ positive i-edges and $2n_b$ negative i-edges. (As always, an edge with both endpoints labeled i is counted twice.)

If i is not a label of a Scharlemann bigon in Γ_b then the $2n_b$ positive i-edges in Γ_b are mutually nonparallel, so $\hat{\Gamma}_b$ has at least $2n_b$ positive edges. By Lemma 2.5 the reduced graph $\hat{\Gamma}_b$ has no more than $3n_b$ edges, so it has at most n_b negative edges. On the other hand, by Lemma 2.7(1) each family of parallel negative edges in Γ_b has at most two endpoints labeled i; since there are $2n_b$ such endpoints, Γ_b must have at least n_b families of negative edges. It follows that Γ_b has exactly n_b families of negative edges, each having exactly two endpoints labeled i.

Suppose $n_b = 6$. Then there are 12 edges in the three families incident to u_i, and by Lemma 2.3(3) each family contains at most four edges, hence each family contains exactly four edges. If some of these edges are loops, then there are four loops and four non-loop edges. No loop can be parallel to a non-loop edge in Γ_b since otherwise the label i would appear three times among a set of parallel edges in Γ_b. It follows that all the 8 positive edges incident to u_i are mutually nonparallel in Γ_b, so the reduced graph $\hat{\Gamma}_b$ would have at least 8 negative edges, which is a contradiction as we have shown above that $\hat{\Gamma}_b$ has exactly $n_b = 6$ negative edges. Hence we can assume there is no loop based at u_i. Note that a family of four parallel edges in Γ_a contains a Scharlemann bigon. If the label pair of the Scharlemann bigon is $\{j, j+1\}$, then these two labels appear twice among the endpoints of this family, and each of the other four labels appears exactly once. By Lemma 2.3(4) at most four labels are the labels of some Scharlemann bigons in Γ_b, so there is some k which is not a label of a Scharlemann bigon and hence appears exactly three times among the endpoints of the positive edges incident to u_i. Dually, this implies that some negative edge in $\hat{\Gamma}_b$ contains only one i-edge, which is a contradiction as we have shown above that each negative edge in $\hat{\Gamma}_b$ must contain exactly two negative i-edges.

The proof for $n_b = 8$ is similar. In this case the numbers of edges in the three positive families incident to u_i are either $(6, 5, 5)$ or $(6, 6, 4)$. Using the fact that there are at most four labels of Scharlemann cycles one can show that in either case some label appears three times among the endpoints of these edges, which would lead to a contradiction as above. $\square$

LEMMA 4.4. *No vertex u_i is incident to at most one edge in $\hat{\Gamma}_a^+$.*

PROOF. By Lemma 3.2 there are no interior vertices, hence by Lemma 2.11 either (i) $\hat{\Gamma}_b^+$ has a circle component, or (ii) $\hat{\Gamma}_b^+$ has a boundary vertex of valence at most 3, or (iii) all vertices of $\hat{\Gamma}_b^+$ are boundary vertices of valence 4.

In case (i) a vertex v_j on the circle component is incident to at most two positive edges with label i at v_j, hence dually there are at most two negative edges with label j at u_i, and hence at least $\Delta - 2 \geq 2$ positive edges with label j at u_i, which is impossible because u_i is incident to at most one family of positive edges and by Lemma 2.3(3) such a family contains at most one edge with label j at u_i.

The proof for case (ii) is similar because by Lemma 2.7(3) a valence 3 boundary vertex v_j of $\hat{\Gamma}_b^+$ is incident to at most $2n_a$ positive edges of Γ_b and hence at most two positive edges with label i at v_j.

In case (iii), since u_i is incident to at most $n_b/2 + 2 < n_b$ positive edges, there is a label j such that all four edges with label j at u_i are negative. Dually v_j has four positive i-edges. Since it is a boundary vertex, it is incident to at least $3n_a + 1$ positive edges. On the other hand, since v_j has valence 4 in $\hat{\Gamma}_b^+$, by Lemma 2.7(2) it has at most $2(n_a + 2) < 3n_a$ positive edges, a contradiction. $\square$

COROLLARY 4.5. *Each component of $\hat{\Gamma}_a^+$ is contained in an essential annulus but not a disk on $\hat{F}_a$.*

PROOF. By Lemma 2.6 $\hat{\Gamma}_a^+$ has at least two components, so if the result is not true then one can find a disk D on $\hat{F}_a$ such that $D \cap \hat{\Gamma}_a^+$ is a component G of $\hat{\Gamma}_a^+$. By Lemma 4.4 G is not an arc, so by Lemma 2.9 it has at least three boundary vertices of valence at most 3. By Lemma 4.3 these vertices are labels of Scharlemann cycles in Γ_b, which is a contradiction because by Lemma 2.3(4) Γ_a contains at most two labels of Scharlemann cycles of each sign. $\square$

Let G be a component of $\hat{\Gamma}_a^+$ contained in the interior of an essential annulus A on $\hat{F}_a$. By Corollary 4.5, G is not contained in a disk, hence it contains some cycles which are topologically essential simple closed curves on $\hat{F}_a$, and all such cycles are isotopic to the core of A. We call such a cycle an *essential cycle* on G. Note that a cycle may have more than two edges incident to a vertex, but an essential cycle does not. An essential cycle C of G is *outermost on A* if all essential cycles of G lie in one component of $A|C$. By cutting and pasting one can see that outermost essential cycles always exist, and there are at most two of them, which we denote by C_l and C_r, called the *leftmost cycle* and the *rightmost cycle*, respectively. Let A_l^l and A_l^r be the components of $A|C_l$, called the *left annulus* and the *right annulus* of C_l, respectively, labeled so that A_l^l contains no essential cycles of G other than C_l. Similarly for A_r^l and A_r^r, where the right annulus A_r^r of C_r is the one that contains no essential cycles other than C_r.

LEMMA 4.6. *The interiors of A_l^l and A_r^r do not intersect G.*

PROOF. Assuming the contrary, let G' be the closure of a component of $G \cap A_l^l$. Since G is connected, G' must intersect C_l at some vertex v, but it cannot intersect C_l at more than one vertex, as otherwise the union of an arc in G' and an arc on C_l would be an essential cycle in A_l^l other than C_l, contradicting the definitions of leftmost cycle and its left annulus. For the same reason, G' contains no essential cycles, hence it lies on a disk D in A_l^l. By Lemma 4.4 G has no vertex of valence 1, so G' is not homeomorphic to an arc. By Lemma 2.9 G' has at least three boundary vertices of valence at most 3. Let v^1 and v^2 be such vertices other than v. They are boundary vertices of G lying in the interior of A_l^l with valence at most 3, and $v^i \neq v$.

By Lemma 4.3, for $i = 1, 2$ there is a Scharlemann bigon $\{e_1^i, e_2^i\}$ on Γ_b with v^i as a label, and by Lemma 2.2(5) $C_i = e_1^i \cup e_2^i$ is an essential curve on $\hat{F}_a$ containing v^i. Since v^2 is a boundary vertex of G', it is not a cut vertex, hence there is an arc C' on G' connecting v^1 to v which is disjoint from v^2. Now the union $C_1 \cup C' \cup C_l$ cuts $\hat{F}_a$ into an annulus and a disk D containing v^2 in its interior, so the cycle C_2 is also contained in the disk D, which is a contradiction to the fact that C_2 is topologically an essential curve on $\hat{F}_a$. □

Lemma 4.6 shows that G is contained in the region R between C_l and C_r. Since G has no interior vertices, all its vertices are on $C_l \cup C_r$. If C_l is disjoint from C_r then R is an annulus, and if $C_l = C_r$ then $R = C_l = C_r$ is a circle. In the generic case we have $C_l \cap C_r = E_1 \cup ... \cup E_k$, where each E_i is either a vertex or an arc. The region R is then a union of these E_i and some disks $D_1, ..., D_k$, such that ∂D_i is the union of two arcs, one in each of C_r and C_l. When $k = 1$ and $E_1 = v$ is a vertex, D_1 is a disk with a pair of boundary points identified to the single point v. Note that a vertex of G is a boundary vertex if and only if it is on $C_l \cup C_r - C_l \cap C_r$.

LEMMA 4.7. *Let $C = C_l$ or C_r.*

(1) If C has a boundary vertex u_i of valence at most 3 then it has no other boundary vertex of valence at most 4.

(2) If C has a boundary vertex u_i of valence 2 then it has no other boundary vertex.

PROOF. (1) By Lemma 4.3, i is a label of a Scharlemann bigon in Γ_b. On Γ_a the edges of this Scharlemann bigon form a cycle C' containing u_i and another vertex u_k. By Lemma 2.2(5) C' is topologically an essential circle on the torus $\hat{F}_a$. Since u_i is a boundary vertex, one can see that C' is topologically isotopic to C. By Lemma 4.6 applied to G and to the component of $\hat{\Gamma}_a^+$ containing u_k, there are no other vertices of Γ_a between C' and C. Hence any boundary vertex $u_j \neq u_i$ on C is incident to at at most one family of parallel negative edges, connecting it to u_k. The result now follows from Lemma 4.2.

(2) Note that since u_i is a boundary vertex, the edges of any Scharlemann bigon on Γ_b with i as a label must connect u_i to the same vertex u_k on Γ_a, so there are at most n_b such bigons because there are only two edges on $\hat{\Gamma}_a$ connecting u_i to u_k, each representing a family of at most n_b edges. By Lemma 2.7(2) u_i is incident to at most $n_b + 2$ positive edges, hence at least $3n_b - 2$ negative edges. If a pair of these edges are parallel on Γ_b then they form a Scharlemann bigon. Hence by the above we see that there are at most n_b pairs of such edges. It follows that $\hat{\Gamma}_b$ has at least $3n_b - 2 - n_b = 2n_b - 2$ positive edges.

If u_j is a boundary vertex of C other than u_i then as in the proof of (1) it is incident to at most one family of negative edges, so it has at least $3n_b$ positive edges. Since no three of those are parallel on Γ_b, we see that $\hat{\Gamma}_b$ has at least $3n_b/2$ negative edges, so $\hat{\Gamma}_b$ would have a total of at least $2n_b - 2 + 3n_b/2 > 3n_b$ edges, contradicting Lemma 2.5. □

Now suppose $C_l \cap C_r \neq \emptyset$, and $C_l \neq C_r$. Then the region R between C_l and C_r can be cut along vertices of $C_l \cap C_r$ to obtain a set of disks, and possibly some arcs. Let D be such a disk. If $C_l \cap C_r$ is a single vertex v then D is obtained by cutting R along v, in which case we use $D \cap G$ to denote the graph on D obtained by cutting G along v.

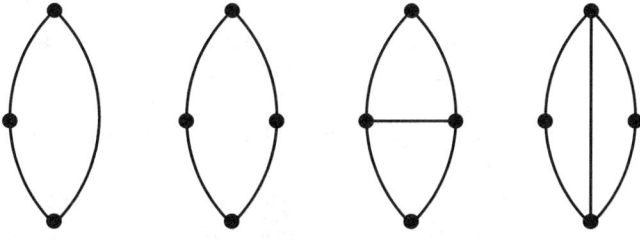

Figure 4.1

LEMMA 4.8. $G' = D \cap G$ is one of the four graphs in Figure 4.1.

PROOF. Let v', v'' be the vertices of G' lying on both C_l and C_r. (Note that they are distinct vertices on G' but may be identified to a single vertex on G.) These vertices divide ∂D into two arcs E_1 and E_2, with $E_1 \subset C_l$ and $E_2 \subset C_r$.

By Lemma 4.7, each E_j contains at most one vertex of valence at most 3 in its interior. Therefore, if D contains at most one interior edge then G' has at most four vertices, so it is one of the four graphs in Figure 4.1. We need to show that D cannot have more than one interior edge.

First suppose there is an interior edge e of G' which has both endpoints on E_1. We may choose e to be outermost in the sense that there is an arc E' on E_1 with $\partial E' = \partial e$, and there is no edge of G' inside the disk bounded by $E' \cup e$. Since G' has no parallel edges, there must be a vertex v in the interior of E', which has valence

2. By Lemma 4.7(2), in this case C_l has no other boundary vertices, so E_1 has no vertex other than v in its interior; in particular, e must have its endpoints on v' and v''. This implies that all interior edges have both endpoints on E_2, and by the same argument as above we see that E_2 has exactly one vertex in its interior, and all edges must have endpoints on v' and v''. Since G' has no parallel edges, it can have at most one edge connecting v' to v'', and we are done.

We can now assume that every interior edge of G' has one endpoint in the interior of each E_i. Let G'' be the union of the interior edges. The above implies that G'' cannot have a cycle, so it is a union of several trees with endpoints in the interiors of E_1 and E_2. A vertex of valence 1 in G'' is a vertex of valence 3 in G', and by Lemma 4.7(1) there is at most one such for each E_i. Therefore G'' is a chain, with two vertices of valence 1 and $k \geq 0$ vertices of valence 2, so G' has one vertex of valence 3 on each E_i, and k vertices of valence 4. Note that these are boundary vertices. However, by Lemma 4.7(1), if G has a vertex of valence 3 on C_l then it has no boundary vertex of valence at most 4 on C_l, and similarly for C_r. It follows that $k = 0$, which again implies that G' has only one interior edge. □

LEMMA 4.9. *If G is a component of $\hat{\Gamma}_a$ and $C_l \cap C_r \neq \emptyset$, then G is one of the graphs in Figure 4.2 (1) – (8).*

PROOF. If $C_l = C_r$ then G is a simple cycle, in which case each vertex has valence 2 and hence is a label of a Scharlemann bigon by Lemma 4.3. By Lemma 2.3(4), G has at most two such vertices, hence G is the graph in Figure 4.2(1) or (2).

Suppose $C_l \neq C_r$ and $C_l \cap C_r \neq \emptyset$. We call the endpoints of $C_l \cap C_r$ *breaking points* of G, which cut the region R between C_l and C_r into several disks $D_1, ..., D_k$ and possibly some arcs. By Lemma 4.8 each $G_i = D_i \cap G$ is one of the graphs in Figure 4.1. We say that G_i is of type (j) if it is the graph in Figure 4.1(j). Since G can have at most two boundary vertices of valence at most three, we see that either $k = 1$, or $k = 2$ and both G_i are of type (1).

First assume that $k = 2$ and G_1, G_2 are of type (1). By Lemma 4.7 the two boundary vertices of G_i must be one on each of C_l, C_r. If the component of $C_l \cap C_r$ containing a breaking point v' on G_i is an arc instead of a vertex, then v' would be a vertex of $\hat{\Gamma}_b$ which is incident to three positive edges, two of which are adjacent, in which case by Lemma 4.3 v' is a label of a Scharlemann bigon in Γ_b. Since G contains no more than two Scharlemann bigon labels, this cannot happen. It follows that G is the graph shown in Figure 4.2(6).

We can now assume $k = 1$. For the same reason as above, we see that if G_1 is of type (1), (2) or (3), then G is as shown in Figure 4.2(3), (5) or (4), respectively. If G_1 is of type (4), the breaking vertices may be incident to an edge in $C_l \cap C_r$, so G is the graph in Figure 4.2(7) or (8). □

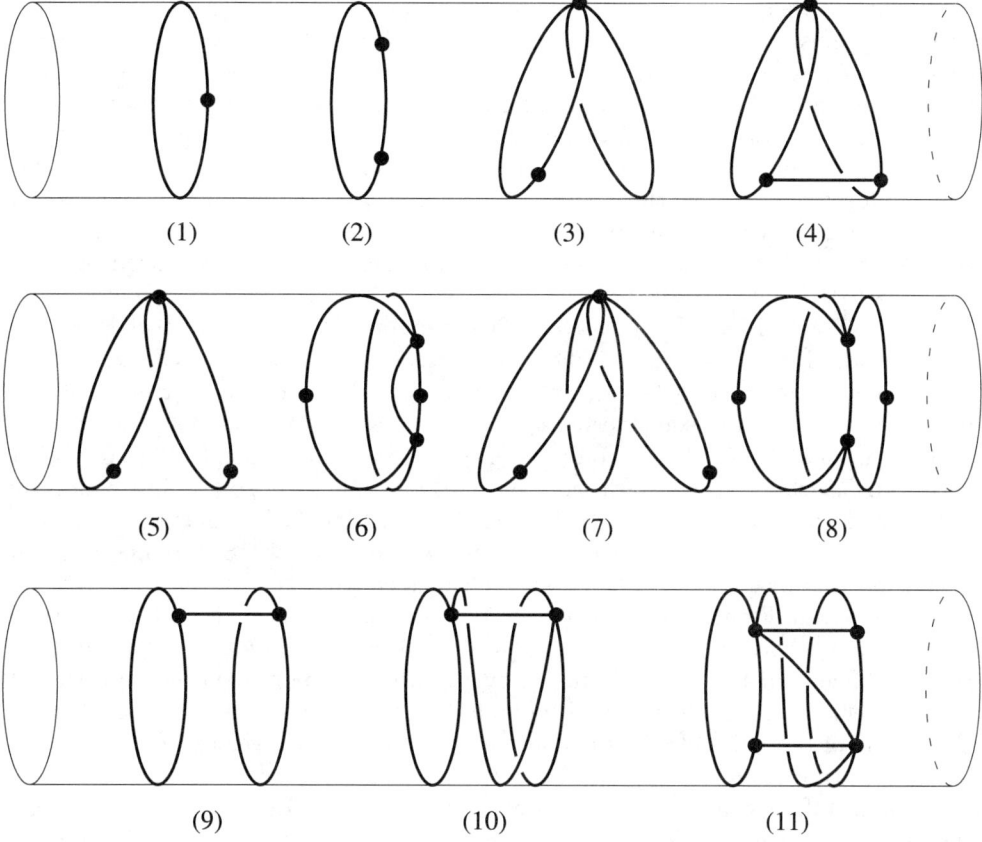

Figure 4.2

LEMMA 4.10. *If G is a component of $\hat{\Gamma}_a^+$ and $C_l \cap C_r = \emptyset$, then G is one of the three graphs in Figure 4.2 (9), (10), or (11).*

PROOF. Note that in this case all vertices on C_l and C_r are boundary vertices. If C_l has a vertex of valence 2 then by Lemma 4.7 it has no other vertices, in which case C_l is a loop and we have $G = C_l$, so $C_l = C_r$, a contradiction. Therefore C_l and C_r have no vertices of valence 2, hence all vertices of G have valence at least 3.

Doubling the annulus and calculating Euler characteristic, we see that

$$\sum (4 - val(v_i)) \geq 0.$$

By Lemma 4.7 G has at most two vertices of valence 3.

First assume that G has two vertices v_1, v_2 of valence 3. By Lemma 4.7 v_1, v_2 cannot both be on C_l or C_r, hence each of C_l and C_r contains exactly one vertex of valence 3. By Lemma 4.7 they cannot contain vertices of valence 4. By the above formula G has either (i) no other vertex, or (ii) one other vertex with valence 6 or 5, or (iii) two other vertices, both having valence 5. One can check that in Case (i) the graph is that of Figure 4.2(9), Case (ii) does not happen, and in Case (iii) the graph is the one in Figure 4.2(11).

If G contains only one vertex v of valence 3, then by the above formula it contains at least one vertex of valence 5, and all other vertices are of valence 4. If

C_l contains v then by Lemma 4.7 it contains no vertices of valence 4. Since each edge of G must have one endpoint on each of C_l and C_r, we see that C_l must contain a vertex of valence 5, and C_r contains exactly two vertices, each of valence 4. One can check that there is no reduced graph satisfying these conditions.

Now assume that G has no vertices of valence 3. Then by the above formula all vertices of G are of valence 4. Since G has no parallel edges, all edges of $G - C_l \cup C_r$ must connect C_r to C_l, so the graph G is completely determined by the number of vertices k on C_r, which must be the same as that on C_l. Denote such a graph by G_k. When $k = 1$, the graph $G = G_1$ is shown in Figure 4.2(10). We need to show that $k > 1$ does not happen.

Suppose $k > 1$. By Lemma 4.2 each vertex on C_l is incident to at least two negative edges in $\hat{\Gamma}_a$. Let G' be the component of $\hat{\Gamma}_a^+$ adjacent to C_l, and let C_r' be the outermost cycle of G' adjacent to C_l. If C_r' has only one vertex then two negative edges based at some vertex v_i on C_l form an essential loop on the annulus between C_r' and C_l, so there is only one negative edge of $\hat{\Gamma}_a$ incident to any other vertex on C_l, which is a contradiction. Similarly if some vertex on C_r' is a boundary vertex of valence at most 3 then by Lemma 4.3 it is a label of a Scharlemann bigon, which is again a contradiction because the two edges of the Scharlemann bigon would form an essential loop as above. These facts rule out the possibility of G' being a graph in Figure 4.2 (1) or (3) – (11). If G' is the one in Figure 4.2(2) then by Lemma 4.3 both of its vertices are labels of Scharlemann bigons. The edges of these two Scharlemann bigons form two cycles, which cannot be on the same side of G' as otherwise one of them would lie on a disk, which contradicts Lemma 2.2(5). Hence one of the pairs of edges connect a vertex of G' to a vertex of C_l, which is again a contradiction.

It now follows that if some component of $\hat{\Gamma}_a^+$ is a G_k for $k \geq 2$, then so are all the other components. Moreover, none of the vertices is a label of a Scharlemann cycle as otherwise some vertex would be incident to a single negative edge in $\hat{\Gamma}_a$, which would contradict Lemma 4.2. Hence Γ_b has no Scharlemann cycles. On the other hand, by Lemma 2.7(2) the four families of positive edges at a vertex v_i of G_k contain at most $2n_b + 4$ edges, so v_i is incident to at least $\Delta n_b - (2n_b + 2) > n_b$ negative edges. By Lemma 2.8 this implies that Γ_b does have a Scharlemann cycle, which is a contradiction. $\square$

COROLLARY 4.11. *Suppose $\Delta \geq 4$, and $n_a > 4$ for $a = 1, 2$. Then*

(1) each component of $\hat{\Gamma}_a^+$ is one of the 11 graphs in Figure 4.2; and

(2) each Γ_a contains a Scharlemann cycle, hence $\hat{F}_b$ is separating, and n_b is even for $b = 1, 2$.

PROOF. (1) This follows from Lemmas 4.9 and 4.10.

(2) By (1), $\hat{\Gamma}_a^+$ contains either a vertex v of valence 2 or a boundary vertex of valence at most 4. In the first case the result follows from Lemma 4.3. In the second case by Lemma 2.7(2) v is incident to at most $2n_b + 4$ positive edges, hence at least $2n_b - 4 > n_b$ negative edges, so by Lemma 2.8 Γ_a has a Scharlemann cycle. $\square$

5. The case $n_1, n_2 > 4$

In this section we will complete the proof that the generic case $n_1, n_2 > 4$ cannot happen. We assume throughout the rest of the section that $n_1, n_2 > 4$. Let G be a component of $\hat{\Gamma}_a^+$. By Corollary 4.11 G is one of the graphs in Figure 4.2.

5. THE CASE $n_1, n_2 > 4$

We need to rule out all these possibilities. Recall that a component of $\hat{\Gamma}_a^+$ is of type (k) if it is the graph in Figure 4.2(k).

Here is a sketch of the proof. We first show (Lemma 5.4) that $\hat{\Gamma}_a^+$ cannot have two boundary vertices of valence 2, hence no component of $\hat{\Gamma}_a^+$ is of type (5)–(8). Types (3) and (11) will be ruled out in Lemmas 5.6 and 5.7, so we are left with types (1), (2), (4), (9) and (10). Lemma 5.8 will show that each vertex of a type (10) component is a label of Scharlemann cycle, which implies that all vertices of $\hat{\Gamma}_a^+$ are labels of Scharlemann cycles, except the valence 4 vertex in a type (4) component. Since $\hat{\Gamma}_a^+$ has at most two Scharlemann labels of each sign, we see that each $\hat{\Gamma}_a^+$ is a union of two type (4) components. This will be ruled out in Lemma 5.10, completing the proof of the theorem.

Each vertex u_i in Γ_a has Δ edge endpoints labeled j. Define $\sigma(u_i, v_j)$ to be the number of those on positive edges minus the number of those on negative edges. In other words, it is the sum of the signs of the edges with an endpoint labeled j at u_i.

Define a vertex u of $\hat{\Gamma}_a^+$ to be *small* if it is either of valence 2 or is a boundary vertex of valence 3. Note that a component of type (1) or (3) in Figure 4.2 has one small vertex, a component of type (10) has no small vertex, and all others have two small vertices.

LEMMA 5.1. *(1)* $\sigma(u_i, v_j) = -\sigma(v_j, u_i)$.
(2) If v_j is a small vertex in $\hat{\Gamma}_b^+$ then $\sigma(u_i, v_j) \geq 0$ for all i.
(3) If $\hat{\Gamma}_a^+$ has a boundary vertex u_i of valence 2, then $\sigma(u_i, v_j) < 0$ for all but at most two j, at most one for each sign.
(4) If $\hat{\Gamma}_a^+$ has a boundary vertex of valence 2, then $\hat{\Gamma}_b^+$ has at most one small vertex of each sign.

PROOF. (1) This follows from the parity rule Lemma 2.2(1).

(2) If v_j has valence 2 in $\hat{\Gamma}_b^+$ then each label i appears at most twice among the positive edge endpoints. If v_j is a boundary vertex of valence 3 in $\hat{\Gamma}_b^+$ then by Lemma 2.7(3) it is incident to at most $2n_a$ adjacent positive edges in Γ_b, hence again each i appears at most twice among the positive edge endpoints. Since $\Delta \geq 4$, the result follows.

(3) If u_i is a boundary vertex of valence 2 then by Lemma 2.7(2) there are at most $n_b + 2$ adjacent positive edges, so at most two labels appear more than once among the positive edge endpoints, and if there are two then they are adjacent, so there is only one for each sign.

(4) This follows immediately from (2) and (3). □

LEMMA 5.2. *Suppose $\hat{\Gamma}_a^+$ has a boundary vertex u_i of valence 2. Then all components of $\hat{\Gamma}_b^+$ are of type (1), (3) or (10). Moreover, for each sign there is at most one component with vertices of that sign which is of type (1) or (3).*

PROOF. This follows immediately from Lemma 5.1(4) and the fact that a component of type (1) or (3) has one small vertex, a component of type (10) has no small vertex, and all others have two small vertices. □

LEMMA 5.3. *Let v_j be a vertex of a type (10) component G of $\hat{\Gamma}_b^+$.*
(1) v_j is incident to at most $2n_a + 2$ positive edges in Γ_b.
(2) $\sigma(u_i, v_j) \geq 0$ for all but at most two u_i, one for each sign.

PROOF. (1) By Lemma 2.7(2) the four families of adjacent parallel positive edges incident to v_j contain $m \leq 2(n_a + 2)$ edges. If $m > 2n_a + 2$, then in particular one of the families contains more than $n_a/2$ edges, so it contains a Scharlemann bigon. By Lemma 2.2(4) and Lemma 2.2(1) the labels at the endpoints of a loop at v_j must have different parity, which rules out the possibility $m = 2n_a + 3$. Hence $m = 2n_a + 4$. Note that in this case there are at least 4 parallel loops $\{e_1, ..., e_4\}$, where e_1 is the outermost edge on the annulus containing G. By looking at the labels at the endpoints of these loops, we see that e_2, e_3 form a Scharlemann bigon, which contradicts Lemma 2.2(6) because $\{e_1, e_4\}$ is then an extended Scharlemann cycle.

(2) Since v_j is a boundary vertex of G, the positive edges incident to v_j are adjacent. Therefore (1) implies that $\sigma(v_j, u_i) \leq 0$ for all but at most two i, hence by Lemma 5.1(1) we have $\sigma(u_i, v_j) \geq 0$ for all but at most two u_i, and if there are two such u_i then they are of opposite sign. $\square$

LEMMA 5.4. $\hat{\Gamma}_a^+$ cannot have two parallel boundary vertices of valence 2; in particular, no component G of $\hat{\Gamma}_a^+$ is of type (5), (6), (7) or (8).

PROOF. Suppose to the contrary that $\hat{\Gamma}_a^+$ has two boundary vertices u_{i_1}, u_{i_2} of valence 2, and of the same sign. By Lemma 5.2, each component G' of $\hat{\Gamma}_b^+$ is of type (1), (3) or (10). If G' is of type (3) then it has a boundary vertex of valence 2, so applying Lemma 5.2 to this vertex (with $\hat{\Gamma}_a^+$ and $\hat{\Gamma}_b^+$ switched), we see that G must be of type (1), (3) or (10), which is a contradiction. Therefore G' must be of type (1) or (10).

By Lemma 5.1(3), $\sigma(u_{i_1}, v_k) < 0$ for all but at most two v_k. Similarly for $\sigma(u_{i_2}, v_k)$. Since $n_b > 4$, there is a vertex v' such that $\sigma(u_r, v') < 0$ for both $r = i_1, i_2$. On the other hand, if v' is on a component G' and if G' is of type (1) then by Lemma 5.1(2) we have $\sigma(u_r, v') \geq 0$ for all u_r, while if G' is of type (10) then Lemma 5.3(2) says $\sigma(u_r, v') \geq 0$ for either $r = i_1$ or i_2 because u_{i_1} and u_{i_2} are of the same sign. This is a contradiction. $\square$

Note that a vertex u on a component G of $\hat{\Gamma}_a^+$ is a boundary vertex if it lies on one outermost essential cycle C_1 of G but not the other one. In this case there is a unique component G' of $\hat{\Gamma}_a^+$ and a unique outermost essential cycle C_2 on G' such that $C_1 \cup C_2$ bounds an annulus on $\hat{F}_a$ whose interior contains no vertex of Γ_a. We say that G' and C_2 are *adjacent* to u.

LEMMA 5.5. Let u_i be a vertex on a type (10) component G of $\hat{\Gamma}_a^+$. If u_i is not a label of a Scharlemann cycle in Γ_b, then
 (i) the component G' of $\hat{\Gamma}_a^+$ adjacent to u_i is of type (1), (3) or (10);
 (ii) u_i is incident to exactly $2n_b - 2$ negative edges; and
 (iii) $\hat{\Gamma}_b^+$ has only two components, each of type (4) or (11).

PROOF. We assume that u_i is not a label of a Scharlemann cycle. Let G' and C be the component and outermost cycle adjacent to u_i. If C has a boundary vertex u_j of valence at most 3, then by Lemma 4.3 u_j is a label of a Scharlemann cycle. Since u_j is a boundary vertex and there is no vertex between C and the outermost cycle on G containing u_i, the edges of the above Scharlemann cycle must connect u_j to u_i, hence u_i is also a label of the Scharlemann cycle, which is a contradiction. Also, if G' is of type (2) then by Lemma 4.1 each of its vertices is a label of a Scharlemann cycle. Recall that the edges of a Scharlemann cycle in Γ_b cannot lie in a disk on $\hat{F}_a$, hence the edges of one of the Scharlemann cycles must connect a vertex on C to u_i, which again is a contradiction. Therefore C does not have a boundary vertex of valence at most 3, and it is not on a type (2) component. Examining the graphs in Figure 4.2, we see that G' must be of type (1), (3) or (10). Moreover, if it is of type (3) then C is the loop there. In any case, C contains only one vertex.

Let t be the number of negative edges incident to u_i. Since C has only one vertex u_j, u_i is incident to at most two families of negative edges $\hat{e}_1, \hat{e}_2$, all connecting u_i to u_j, so by Lemma 2.7(1) $t \leq 2n_b$. On the other hand, by Lemma 5.3 u_i is incident to at most $2n_b + 2$ positive edges, so $t \geq 2n_b - 2$. Therefore we have $2n_b \geq t \geq 2n_b - 2$.

First assume $t = 2n_b$. Then each of $\hat{e}_1$ and $\hat{e}_2$ contains exactly n_b edges. Since u_i is not a label of Scharlemann cycle, by Lemma 2.4 these $2n_b$ edges are mutually non-parallel on Γ_b, hence $\hat{\Gamma}_b^+$ has at least $2n_b$ edges. On the other hand, by Lemma 2.10(2) it cannot have more than $2n_b$ such edges, hence $\hat{\Gamma}_b^+$ has exactly $2n_b$ edges, each containing exactly one edge in $\hat{e}_1 \cup \hat{e}_2$. Counting the number of edges on each graph in Figure 4.2, we see that each component of $\hat{\Gamma}_b^+$ must be of type (10) or (11). Also, a component of type (11) has a vertex v_k of valence 5 in $\hat{\Gamma}_b^+$, so the above implies that the label k appears 5 times among the endpoints of edges in $\hat{e}_1 \cup \hat{e}_2$, which is absurd. This rules out the possibility for a component to be of type (11). Now notice that these two families of n_b parallel edges have the same transition function, hence if some edge has the same labels on its two endpoints, then they all do. It follows that no component can be of type (10) because it has both loop and non-loop edges. This completes the proof for the case $t = 2n_b$.

If $t = 2n_b - 1$ then one of $\hat{e}_1, \hat{e}_2$ contains n_b edges and the other contains $n_b - 1$ edges. Examining the labels at the endpoints of these edges we see that if an edge in $\hat{e}_1$ has labels of the same parity at its two endpoints then an edges in $\hat{e}_2$ would have labels of different parities at its endpoints, and vice versa. This contradicts the parity rule (Lemma 2.2(1)).

We can now assume $t = 2n_b - 2$. Without loss of generality we may assume that the labels of the endpoints of $\hat{e}_1 \cup \hat{e}_2$ appear as $1, 2, ..., n_b, 1, ..., n_b - 2$ on ∂u_i when traveling clockwise, and we assume that the first n_b are endpoints of $\hat{e}_1$. (The other cases are similar.) Let e_p^k ($k = 1, 2$) be the edge in $\hat{e}_k$ with label p at u_i, and assume that the label of e_1^1 on u_j is $1 + r$ for some r. Then one can check that the label of e_p^1 on u_j is $p + r$, and the label of e_p^2 on u_j is $p + r + 2$. (All labels are integers mod n_b.) Hence for any p between 3 and n_b, the edges e_p^1 and e_{p-2}^2 have the same label $p + r$ at u_j. On Γ_b this implies that there are two positive edges, connecting v_p to v_{p+r} and v_{p+r} to v_{p-2}, so v_p are v_{p-2} are in the same component of $\hat{\Gamma}_b^+$. Since this is true for all p between 3 and n_b, it follows that $\hat{\Gamma}_b^+$ has only two components.

By Lemmas 2.8 and 2.2(4) $\hat{\Gamma}_b^+$ has the same number of positive vertices and negative vertices, hence each component G has at least three vertices. This rules out the possibility for G to be of type (1), (2), (3), (9) or (10). Combined with Lemme 5.4 we see that each component of $\hat{\Gamma}_b^+$ is of type (4) or (11). □

LEMMA 5.6. *No component of $\hat{\Gamma}_a^+$ is of type (3).*

PROOF. By Lemma 5.2 if $\hat{\Gamma}_a^+$ has a component of type (3) then each component of $\hat{\Gamma}_b^+$ is of type (1), (3) or (10), and there is at most one component of type (1) or (3) for each sign. Since $n_b > 4$ and a component of type (1) or (3) has at most

2 vertices, there is at least one component G of $\hat{\Gamma}_b^+$ of type (10) and at least one other component G' of the same sign. On the other hand, by Lemma 5.5 each vertex of G is a label of a Scharlemann cycle, and by Lemmas 4.1 and 4.3 at least one vertex of G' is a label of a Scharlemann cycle, so there are at least three labels of Scharlemann cycles of the same sign, contradicting Lemma 2.3(4). □

LEMMA 5.7. *No component of $\hat{\Gamma}_a^+$ is of type (11).*

PROOF. An outermost cycle on a component G of type (11) contains two parallel vertices u_i and u_j, where u_i is of valence 3 and hence the label of a Scharlemann bigon (Lemma 4.3), and u_j has valence 5. If $\{e_1, e_2\}$ is a Scharlemann bigon on Γ_b with label pair $\{i, i+1\}$, say, then on Γ_a these edges form an essential curve containing the vertices u_i and u_{i+1}, which separates u_j from all other vertices of opposite sign, hence all negative edges incident to u_j have their other endpoints on u_{i+1}, and they are all parallel. Thus u_j has at most n_b negative edges, and hence at least $3n_b$ adjacent positive edges. In particular, each label appears at least three times among endpoints of positive edges at u_j. Dually, each vertex v_k in Γ_b is incident to at least three negative edges labeled j at v_k. If v_k is a boundary vertex, then this implies that it is incident to at least $2n_a + 1$ negative edges, so by Lemma 4.1 it is a label of a Scharlemann cycle.

By Lemmas 5.4 and 5.6 a component of $\hat{\Gamma}_b^+$ is of type (1), (2), (4), (9), (10) or (11). By the above and Lemma 4.1 all vertices of Γ_b except those with valence 4 in type (4) components are labels of Scharlemann cycles. Since $n_b > 4$ and there are at most two Scharlemann labels for each sign, we see that $\hat{\Gamma}_b^+$ has only two components, each of type (4), so $n_b = 6$, and $\hat{\Gamma}_b^+$ has 10 positive edges. By Lemma 2.5 $\hat{\Gamma}_b$ has at most $3n_b - 10 = 8$ negative edges. On the other hand, we have shown that u_j in Γ_a is incident to at least $3n_b = 18$ positive edges; since no three of them are parallel in Γ_b, $\hat{\Gamma}_b$ has at least $18/2 = 9$ negative edges, which is a contradiction. □

LEMMA 5.8. *Each vertex of a type (10) component of $\hat{\Gamma}_a^+$ is a label of a Scharlemann bigon.*

PROOF. Suppose that a vertex u_i of a type (10) component of $\hat{\Gamma}_a^+$ is not a label of a Scharlemann bigon. By Lemmas 5.5 and 5.7 $\hat{\Gamma}_b^+$ is a union of two type (4) components, so $n_b = 6$, $\hat{\Gamma}_b$ has 10 positive edges, and no more than $3n_b - 10 = 8$ negative edges.

By Lemma 5.5(ii) u_i is incident to $(\Delta - 2)n_b + 2 = 6\Delta - 10$ positive edges (loops counted twice). By Lemma 2.7(1) no three of these are parallel in Γ_b, hence they represent at least $3\Delta - 5$ negative edges in $\hat{\Gamma}_b$. Therefore $\Delta = 4$, and we have at least 7 negative edges in $\hat{\Gamma}_b$. We need to find two more to get a contradiction.

By Lemma 5.5(i) and Lemma 5.6 the component G of $\hat{\Gamma}_a^+$ adjacent to u_i is of type (1) or (10), so the outermost cycle of G adjacent to u_i has a single vertex u_j and a single edge E_0. We claim that E_0 contains at least two edges of Γ_b.

If G is of type (10), then u_j is incident to four families of positive edges in Γ_a, with a total of $2n_b + 2 = 14$ edges, where loops are counted twice. By Lemma 2.3(3) each family contains no more than 4 edges, so the loop edge E_0 contains at least $(14 - 2 \times 4)/2 = 3$ edges of Γ_b. If G is of type (1) then since no three negative edges incident to u_j are parallel in Γ_b, and since $\hat{\Gamma}_b^+$ has only 10 edges, we see that u_j is incident to at most 20 negative edges, hence E_0 contains at least $(24-20)/2 = 2$ edges. This completes the proof of the above claim.

Let e_1', e_2' be the two edges in E_0 closest to u_i. By Lemma 2.2(2) they are not parallel on Γ_b. We claim that on Γ_b neither of them is parallel to any edge incident to u_i, hence $\hat{\Gamma}_b$ contains at least $7 + 2 = 9$ negative edges. This will be a contradiction as we have shown above that $\hat{\Gamma}_b$ has at most 8 negative edges.

By Lemma 5.5(ii) there are exactly $2n_b - 2 = 10$ negative edges $e_1, ..., e_{10}$ connecting u_i to u_j. Without loss of generality we may assume that the sequence of labels of the endpoints of these edges at u_i is $1, ..., 6, 1, ..., 4$, counting clockwise, and the labels of their endpoints at u_j are $r+2, r+3, ..., r-1$, counting counterclockwise. Thus $\{e_1', e_2'\}$ is a Scharlemann bigon with label pair $\{r, r+1\}$.

Since e_i' is a loop, by Lemma 2.3(5) if it is parallel in Γ_b to an edge e incident to u_i then e is also a loop. Note that e_i' and e must have the same label pair. Let E_3 be the loop of $\hat{\Gamma}_a$ based at u_i. It has at most four edges $e_1'', e_2'', e_3'', e_4''$, with label pairs $\{5,6\}, \{6,5\}, \{1,4\}, \{2,3\}$, respectively. By Lemma 2.3(2) we have $\{r, r+1\} \neq \{2, 3\}$, hence if e_i' is parallel to some e_j'' then $\{r, r+1\} = \{5, 6\}$, so $r = 5$, and hence the label sequence of the above negative edges at u_j is also $1, ..., 6, 1, ..., 4$.

The 10 edges $e_1, ..., e_{10}$ are divided into two families E_1, E_2. Since $|E_i| \leq 6$, we have $|E_1| = 4, 5,$ or 6. If $|E_1| = 5$ then the edge e_1 would have label 1 at u_i and label 6 at u_j. Since v_1 and v_6 on Γ_b are antiparallel, this is impossible by the parity rule. If $|E_1| = 4$ then e_1 has the same label 1 at its two endpoints, which contradicts the fact that Γ_b has no loop. Similarly if $|E_1| = 6$ then e_7 has the same label 1 at its two endpoints, which is again a contradiction. This completes the proof of the Lemma. □

LEMMA 5.9. *Each $\hat{\Gamma}_a^+$ is a union of two type (4) components.*

PROOF. By Lemmas 5.4, 5.6 and 5.7, each component G of $\hat{\Gamma}_a^+$ is of type (1), (2), (4), (9) or (10). By Lemmas 4.1, 4.3 and 5.8, we see that all vertices u_i of G are labels of Scharlemann bigons, unless G is of type (4) and u_i is the vertex of valence 4 in G. Since $n_a > 4$ and $\hat{\Gamma}_a^+$ has at most two vertices which are labels of Scharlemann bigons for each sign, we see that $\hat{\Gamma}_a^+$ consists of exactly two components, each of type (4). □

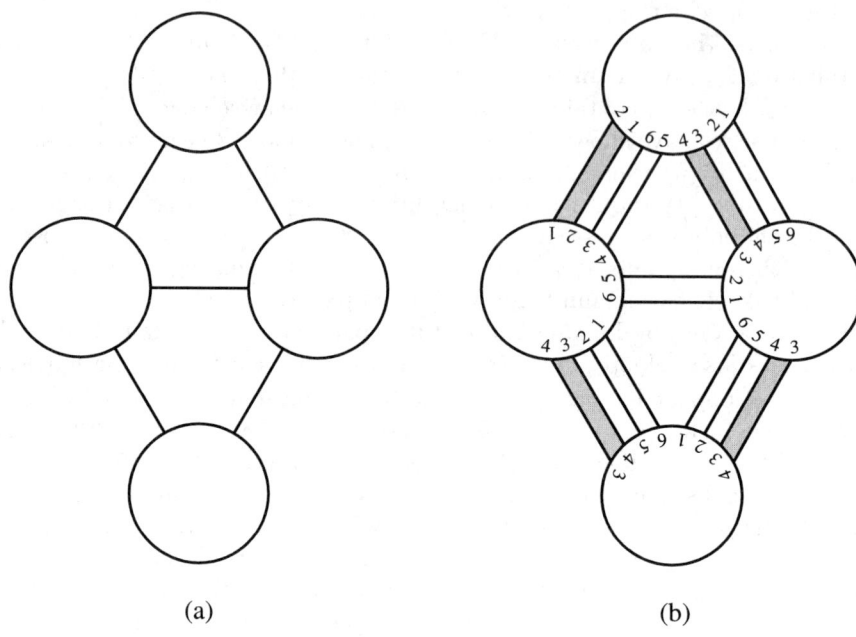

Figure 5.1

LEMMA 5.10. *One of the $\hat{\Gamma}_a^+$ is not a union of two type (4) components.*

PROOF. Assume that each $\hat{\Gamma}_a^+$ is a union of two type (4) components. Each vertex of Γ_a has valence $\Delta n_b \geq 24$, hence Γ_a has at least 72 edges. Since a positive edge in Γ_a is a negative edge in Γ_b, we may assume that Γ_1 has no more negative edges than positive edges, so Γ_1^+ has at least 36 positive edges. Thus one component G of Γ_1^+ has at least 18 edges. Denote by $\hat{G}$ the reduced graph of G. It is of type (4), so it is obtained from the graph in Figure 5.1(a) by identifying the top and bottom vertices.

Let $E_1, ..., E_5$ be the edges of $\hat{G}$. Denote by $|E_i|$ the number of edges of G in E_i, and call it the *weight* of E_i. By Lemma 2.3(3), each $|E_i| \leq 4$. Since G has at least 18 edges, up to relabeling the weights of the edges are at least $(4, 4, 4, 4, 2)$ or $(4, 4, 4, 3, 3)$.

Let D be a triangle face of $\hat{G}$, and let E_1, E_2, E_3 be the edges of D. We will also use D to denote the corresponding triangle face in G. If $|E_i| = 4$ then by Lemma 2.4 E_i contains a Scharlemann bigon, which must be at one end of the family of parallel edges in E_i. We say that the Scharlemann bigon in E_i is *adjacent to* D if one of its edges is on the boundary of D.

5. THE CASE $n_1, n_2 > 4$

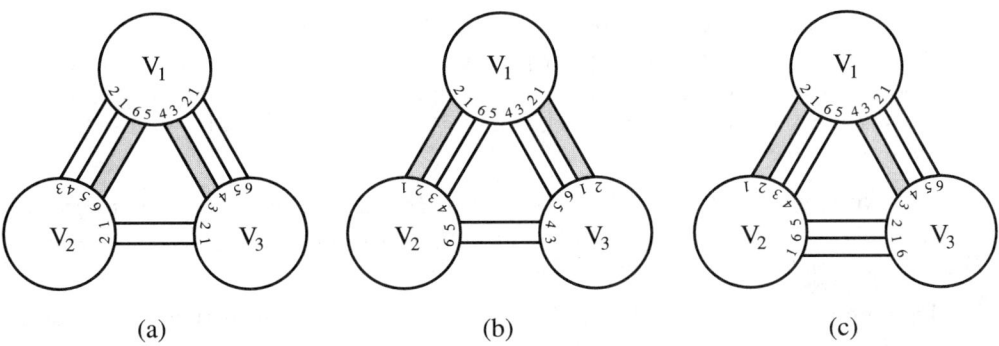

Figure 5.2

Sublemma *If $|E_1| = |E_2| = 4$, then (i) $|E_3| = 2$, and (ii) exactly one of E_1 and E_2 has its Scharlemann bigon adjacent to D.*

PROOF. Let V_1 be the fat vertex incident to both E_1 and E_2. Without loss of generality we may assume that the labels on ∂V_1 are as shown in Figure 5.2(a), where E_1 is the upper right family of edges. Note that the positions of the Scharlemann bigons in E_1, E_2 determine the labels on ∂V_2 and ∂V_3.

If both E_1 and E_2 have their Scharlemann bigons adjacent to D, then the labels are as shown in Figure 5.2(a), in which case we have three Scharlemann bigons with disjoint label pairs, contradicting Lemma 2.3(4). If both Scharlemann bigons of E_1, E_2 are non-adjacent to D, then the labels are as shown in Figure 5.2(b), in which case the edges adjacent to those of D form an extended Scharlemann cycle, which contradicts Lemma 2.2(6). This proves (ii).

We may now assume without loss of generality that the Scharlemann bigon of E_1 is adjacent to D while that of E_2 is not adjacent to D. See Figure 5.2(c). In this case the label pair of the Scharlemann bigon in E_1 is $\{3, 4\}$. If $|E_3| \geq 3$ then E_3 contains a Scharlemann bigon with label pair $\{6, 1\}$. This contradicts Lemma 2.3(2), completing the proof of the sublemma. □

If the weights of the E_i are at least $(4, 4, 4, 3, 3)$, or if the weights are $(4, 4, 4, 4, 2)$ and the horizontal edge in Figure 5.1(a) has weight 4, then the boundary edges of one of the triangles in Figure 5.1(a) have weights $(4, 4, 3)$ or $(4, 4, 4)$, which contradicts the sublemma. Therefore the edges of G are exactly as shown in Figure 5.1(b). As in the proof of the sublemma, we may assume that the labels at the three vertices in the upper triangle of G are as shown in Figure 5.1(b). The Scharlemann bigons in the upper triangle have label pairs $\{3, 4\}$ and $\{1, 2\}$, hence by Lemma 2.3(4) G cannot have a Scharlemann bigon on label pair $\{5, 6\}$. Therefore the labels of the endpoints of the lower-right edges must be as shown in Figure 5.1(b). This determines the labels at the lower vertex. But then neither Scharlemann bigon in the lower triangle is adjacent to the triangle, contradicting the sublemma. □

PROPOSITION 5.11. *The case that both $n_1, n_2 > 4$ is impossible.*

PROOF. This follows from the contradiction between Lemma 5.9 and Lemma 5.10. □

6. Kleinian graphs

In Sections 6 – 11 we will improve Proposition 5.11 to show that $n_i \leq 2$ for $i = 1$ or 2. For the most part we will assume that $n_a = 4$. In this section we prove some useful lemmas. In particular, Lemmas 6.2 – 6.5 study kleinian graphs. Lemma 6.2 gives basic properties of kleinian graphs, which will also be used later in studying the case $n_a = 2$.

DEFINITION 6.1. *The graph Γ_a is said to be kleinian if $\hat{F}_a$ bounds a twisted I-bundle over the Klein bottle $N(K)$ such that each component of $N(K) \cap V_a$ is a $D^2 \times I$, and each component of $N(K) \cap F_b$ is a bigon.*

By Lemma 2.12, if $M(r_a)$ contains a Klein bottle K intersecting K_a at $n_a/2$ points then $\partial N(K)$ is an essential torus intersecting K_a at n_a points, hence in this case we may assume that $\hat{F}_a = \partial N(K)$, where $N(K)$ is a small regular neighborhood of K; in particular, Γ_a is kleinian. In this case $N(K)$ is called the *black region*, and all faces of Γ_b lying in this region are called *black faces*, and the others *white faces*. We assume that the vertices of Γ_a have been labeled so that $u_{2i-1} \cup u_{2i}$ lie on the same component of $V_a \cap N(K)$. The following lemma lists the main properties of kleinian graphs.

LEMMA 6.2. *Suppose Γ_a is kleinian. Then*
(1) each black face of Γ_b is a bigon;
(2) each family of parallel edges in Γ_b contains an even number of edges;
(3) Γ_b has no white Scharlemann disk, hence any Scharlemann cycle of Γ_b has label pair $\{k, k+1\}$ with k odd;
(4) there is a free involution of $\hat{F}_a$, which preserves Γ_a, sending u_{2i-1} to u_{2i} and preserving the labels of edge endpoints.

PROOF. (1) follows from the definition. (2) follows from (1) because if there is a family containing an odd number of edges then one side of that family would be adjacent to a black face, which is not a bigon.

(3) Each edge of a white face is adjacent to a black bigon, so if there is a white Scharlemann disk then the edges of the Scharlemann cycle and the adjacent edges would form an extended Scharlemann cycle, which would be a contradiction to Lemma 2.2(6).

(4) We may assume that the Dehn filling solid torus V_a and the surface F_b intersect $N(K)$ in I-fibers. Thus the involution of $\hat{F}_a$ obtained by mapping each point to the other end of the I-fiber gives rise to the required involution of Γ_a. □

LEMMA 6.3. *Suppose $n_a = 4$. Then Γ_a is kleinian if each vertex of Γ_a is a label of a Scharlemann bigon in Γ_b.*

PROOF. Without loss of generality we may assume that Γ_b has a (12) Scharlemann bigon. By assumption there is a Scharlemann bigon with 3 as a label. If there is no (34) Scharlemann bigon then this Scharlemann bigon must have label pair (23). Similarly the Scharlemann bigon with 4 as a label must have label pair (14). We may therefore relabel the vertices of Γ_a so that the label pairs of the above Scharlemann bigons are (12) and (34) respectively.

Shrinking the Dehn filling solid torus to its core, the Scharlemann bigons become Möbius bands B_{12} and B_{34} in $M(r_a)$. The union of these Möbius bands, together with an annulus on $\hat{F}_a$, becomes a Klein bottle which can be perturbed to intersect the core of the Dehn filling solid torus at $2 = n_a/2$ points. By the convention after Definition 6.1, $\hat{F}_a$ should have been chosen so that Γ_a is kleinian. □

LEMMA 6.4. *Suppose $n_a = 4$. Then Γ_a is kleinian if one of the following holds.*
(1) Γ_b has a family of 4 parallel positive edges.
(2) Γ_b is positive.
(3) $\hat{\Gamma}_b^+$ has a full vertex v_j of valence at most 7.
(4) $\hat{\Gamma}_b^+$ contains 4 adjacent families of positive edges with a total of at least 12 edges.

PROOF. (1) Each label appears exactly twice among the edge endpoints of a family of four parallel positive edges, hence by Lemma 2.4 it is a label of a Scharlemann bigon.

(2) If Γ_b is positive then every vertex u_i of Γ_a is incident to at least $4n_b$ negative edges, two of which must be parallel in Γ_b because by Lemma 2.5 $\hat{\Gamma}_b$ contains at most $3n_b$ edges. Hence by Lemma 2.4 these two edges form a Scharlemann bigon with i as a label. Since this is true for all i, Γ_a is kleinian by Lemma 6.3.

(3) Consider the subgraph G of $\hat{\Gamma}_a$ consisting of negative edges. Then the signs of the vertices around the boundary of a face of G alternate, hence each face has an even number of edges. Using an Euler characteristic argument one can show that G contains at most $2n_a = 8$ edges. By (2) we may assume Γ_b is not positive, so by Lemma 2.3(1) no 3 j-edges are parallel on Γ_a, hence $\Delta = 4$ and G has exactly 8 negative edges, each containing exactly 2 j-edges, with one j label at each ending vertex. Since each vertex u_i has 4 j-labels, we see that u_i is incident to exactly 8 j-edges, two of which must be parallel in Γ_b because $val(v_j, \hat{\Gamma}_b^+) \leq 7$. By Lemma 2.4 they form a Scharlemann bigon with i as one of its labels.

(4) By (1) we may assume that each family contains exactly 3 edges, so the labels at the endpoints of the middle edge in each family are the labels of a Scharlemann bigon. It is easy to see that the 4 endpoints of the middle edges at the vertex are mutually distinct, hence include all labels. □

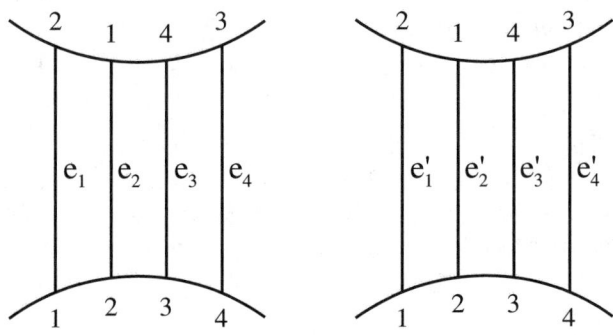

Figure 6.1

LEMMA 6.5. *Suppose $n_a = 4$. Let $e_1 \cup e_2 \cup e_3 \cup e_4$ and $e_1' \cup e_2' \cup e_3' \cup e_4'$ be two families of parallel positive edges in Γ_b as shown in Figure 6.1. Then e_i is parallel to e_i' on Γ_a for all i.*

PROOF. Since $e_1 \cup e_2$ and $e_3 \cup e_4$ form two disjoint essential cycles on $\hat{F}_a$ by Lemma 2.2(5), any (12)-edge must be parallel to e_1 or e_2 and any (34)-edge parallel to e_3 or e_4 on Γ_a. Note also that if e_1 is parallel to e'_2 on Γ_a then e_2 must be parallel to e'_1 (instead of e'_2) on Γ_a as otherwise e_1, e_2 would be parallel on both graphs. Therefore if the result is not true then either e_1 is parallel to e'_2 or e_4 is parallel to e'_3, so there is a subset $e'_r \cup ... \cup e'_s$ of the second family containing less than 4 edges, such that $e'_r \cup e_1$ and $e'_s \cup e_4$ are parallel pairs on Γ_a. This contradicts Lemma 2.19. □

LEMMA 6.6. *Suppose $n_a = 4$ and Γ_b is non-positive.*

(1) No vertex v_j of Γ_b can have two families of 4 positive edges with the same label sequence on ∂v_j. In particular, v_j cannot have two adjacent families of 4 positive edges.

(2) If Γ_a is kleinian, then two adjacent families of positive edges of Γ_b contain at most 6 edges, three contain at most 10, and four contain at most 12.

(3) A full vertex of $\hat{\Gamma}_b^+$ has valence at least 6.

PROOF. (1) If there are two families of 4 positive edges with the same label sequence on ∂v_j then by Lemma 6.5 the two starting edges e_1, e'_1 of these families will be parallel in Γ_a. If e_1, e'_1 have label i at v_j then on Γ_a they have the same label j at u_i, so there are $n_b + 1$ parallel negative edges at u_i, and hence by Lemma 2.3(1) Γ_b would be positive, a contradiction.

(2) By Lemma 6.2(2) the number of edges in each family of positive edges is either 2 or 4, so by (1) two adjacent families contain a total of at most 6 edges. The other two cases follow from this.

(3) Otherwise by Lemma 6.4(3) Γ_a is kleinian, so the weight of each positive family of Γ_b is either 2 or 4. If some full vertex v_i has valence 5 or less in $\hat{\Gamma}_b^+$ then it has two adjacent edge of weight 4, contradicting (1). □

A bigon is called a *non-Scharlemann bigon* if it is not a Scharlemann bigon.

LEMMA 6.7. *Suppose $n_a = 4$ and Γ_a is kleinian.*

(1) Exactly one edge on the boundary of a triangle face of $\hat{\Gamma}_b^+$ represents a non-Scharlemann bigon. Each of the other two represents either a Scharlemann bigon or a union of two Scharlemann bigons.

(2) If some vertex v_i is incident to two edges of weight 4 in $\hat{\Gamma}_b^+$ then any other edge of $\hat{\Gamma}_b^+$ incident to v_i represents a non-Scharlemann bigon.

PROOF. (1) Let $\hat{e}_1, \hat{e}_2, \hat{e}_3$ be the edges of a triangle face δ of $\hat{\Gamma}_b^+$. By Lemma 6.2(2) each edge of $\hat{\Gamma}_b^+$ represents 2 or 4 edges. From the labeling of the edges around δ one can see that there are exactly one or three $\hat{e}_i$ which are neither a Scharlemann bigon nor a union of two Scharlemann bigons. If there are three then they form an extended Scharlemann cycle, which is impossible by Lemma 2.2(6). Hence there must be exactly one such $\hat{e}_i$.

(2) Otherwise v_i would be incident to 5 Scharlemann bigons, three of which have the same label pair, say $\{1, 2\}$. Then on Γ_a there are six i-edges connecting u_1 to u_2, which form at most two families because there is a Scharlemann cocycle containing u_3, u_4. It follows there there are three i labels at the endpoints of a family, so it contains more than n_b edges, contradicting Lemma 2.3(1). □

Suppose $\Delta = 4$. Then a label j is a *jumping label* at u_i if the signs of the four j-edges incident to u_i alternate.

LEMMA 6.8. *Suppose* $\Delta = 4$. *Then a label i is a jumping label at v_j if and only if j is a jumping label at u_i. In particular, if v_j is a boundary vertex of $\hat{\Gamma}_b^+$ then j is not a jumping label at any u_i.*

PROOF. This follows from the Jumping Lemma 2.18. Let $x_1, ..., x_4$ be the four points of $u_i \cap v_j$. Since $\Delta = 4$, the jumping number must be ± 1. Therefore they appear in this order on both ∂u_i and ∂v_j, appropriately oriented. If j is a jumping label at u_i then we may assume x_1, x_3 are positive edge endpoints and x_2, x_4 are negative edge endpoints on ∂u_i, which by the parity rule implies that x_1, x_3 are negative edge endpoints and x_2, x_4 are positive edge endpoints on ∂v_j, hence i is a jumping label at v_j. □

LEMMA 6.9. *Suppose $n_a = 4$, $n_b \geq 4$, and Γ_a is non-positive. Then $\hat{F}_a$ is separating. In particular, u_1 is parallel to u_3 and antiparallel to u_2 and u_4.*

PROOF. The result follows from Lemmas 2.8 and 2.2(4) if some vertex u_i is incident to more than n_b negative edges. In particular, since each family of positive edges contains no more than n_b edges, the result is true if $val(u_i, \hat{\Gamma}_a^+) \leq 2$ for some i. Hence we may assume that $val(u_i, \hat{\Gamma}_a^+) > 2$ for all i. In particular, no component of $\hat{\Gamma}_a^+$ is an isolated vertex. Since $n_a = 4$ and $\hat{\Gamma}_a$ is non-positive, each component G of $\hat{\Gamma}_a^+$ must have exactly two vertices, hence $val(u_i, \hat{\Gamma}_a^+) > 2$ implies that G must be as shown in Figure 4.2(9) or (10). In either case $\hat{\Gamma}_a^+$ has a boundary vertex u_i of valence at most 4, so if $n_b > 4$ then by Lemma 2.7(2) the 4 families of positive edges contain at most $2(n_b + 2) < 3n_b$ edges, hence u_i is incident to more than n_b negative edges and the result follows. Similarly if $n_b = 4$ and some component of $\hat{\Gamma}_a^+$ is of type (9) in Figure 4.2 then by Lemmas 6.4 and 6.6(2) the three positive families at a boundary vertex u_i of valence 3 in $\hat{\Gamma}_a^+$ contain less then 12 edges, hence u_i is incident to more than n_b negative edges and we are done. Therefore we may assume that $n_b = 4$, each component of $\hat{\Gamma}_a^+$ is of type (10) in Figure 4.2, and each u_i is incident to at least 12 positive edges.

In this last case by Lemma 6.4(4) Γ_b is kleinian, so by Lemma 6.2(2) each family of positive edges of Γ_a contains either 2 or 4 edges. Since there is a total of at least 12 positive edges incident to u_i and by Lemma 6.6(2) two adjacent families contain at most 6 edges, the weights of the four edges of $\hat{\Gamma}_a^+$ incident to u_i must be $(4, 2, 4, 2)$ successively. However since the first and the last belong to a loop in $\hat{\Gamma}_a^+$, their weights must be the same, which is a contradiction. □

7. If $n_a = 4$, $n_b \geq 4$ and $\hat{\Gamma}_a^+$ has a small component then Γ_a is kleinian.

A component of $\hat{\Gamma}_a^+$ is *small* if it has at most two edges; otherwise it is *large*. In this section we will show that if $n_a = 4$, $n_b \geq 4$ and $\hat{\Gamma}_a^+$ has a small component then Γ_a is kleinian. It is easy to see that the assumption implies that either $val(u_1, \hat{\Gamma}_a^+) \leq 1$, or $val(u_1, \hat{\Gamma}_a^+) = val(u_3, \hat{\Gamma}_a^+) = 2$ up to relabeling. (See the proof of Proposition 7.6.) The two cases are handled in Lemmas 7.3 and 7.5, respectively.

LEMMA 7.1. *Suppose Γ_a contains a loop edge at u_3. Then Γ_b cannot contain both (12)- and (14)-Scharlemann bigons.*

PROOF. The loop e at u_3 must be essential, otherwise it would bound some disk containing some vertex and hence one of the Scharlemann cocycles in its interior, which contradicts Lemma 2.2(5). Now the (12)- and (14)-Scharlemann bigons in

Γ_b form two essential cycles in Γ_a disjoint from e, so they must be isotopic on $\hat{F}_a$, bounding a disk face containing no vertices of Γ_a in its interior. This is a contradiction to Lemma 2.13. □

LEMMA 7.2. *Suppose $n_a = 4$ and $n_b \geq 4$. If $val(u_1, \hat{\Gamma}_a^+) \leq 1$ and $\hat{\Gamma}_b^+$ has a boundary vertex v_j of valence at most 3, then Γ_a is kleinian.*

PROOF. The assumption $val(u_1, \hat{\Gamma}_a^+) \leq 1$ implies that Γ_a is non-positive, so by Lemma 6.9 u_1 is parallel to u_3 and antiparallel to u_2, u_4. Since u_1 is incident to at most 1 family of positive edges, it is incident to at least three negative j-edges at u_1, so v_j has at least three positive edge endpoints labeled 1. Hence v_j being a boundary vertex implies that it has at least 9 positive edges. If v_j is incident to 10 or more positive edges of Γ_b then it has a family of 4 parallel positive edges and hence Γ_a is kleinian. Therefore we may assume that it has exactly 9 positive edges, divided into three families of parallel edges, each family containing exactly three edges. See Figure 7.1.

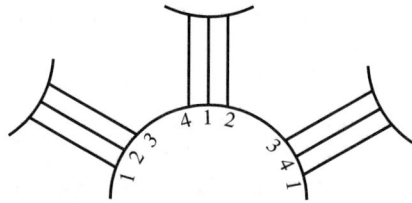

Figure 7.1

Since $n_a = 4$, each of these families contains a Scharlemann bigon, so the labels of the middle edge in the family are labels of a Scharlemann bigon. It follows that $1, 2, 4$ are labels of Scharlemann bigons. Thus if the result is not true then 3 is not a label of Scharlemann bigon, and Γ_b contains both (12)- and (14)-Scharlemann bigons.

There are 7 adjacent negative edges at v_j, so three of them have labels 1 or 3 at v_j. These cannot all be parallel in Γ_a as otherwise there would be three j-edges in a family and hence the family would contain more than n_b edges, contradicting Lemma 2.3(1) and the fact that Γ_b is not positive. On $\hat{\Gamma}_a^+$ this implies that there are at least two edges with endpoints on $\{u_1, u_3\}$, hence $val(u_1, \hat{\Gamma}_a^+) \leq 1$ implies that there is a loop $\hat{e}$ based at u_3. Since Γ_b contains both (12)- and (14)-Scharlemann bigons, this is a contradiction to Lemma 7.1. □

LEMMA 7.3. *Suppose $n_a = 4$ and $n_b \geq 4$. If $val(u_i, \hat{\Gamma}_a^+) \leq 1$ for some i then Γ_a is kleinian.*

PROOF. If Γ_b is positive then Γ_a is kleinian by Lemma 6.4(2). Therefore we may assume that Γ_b is non-positive. By Lemmas 2.3(3) and 2.3(1) each family of parallel edges in Γ_a contains at most n_b edges. Also, notice that since u_i is incident to more than n_b negative edges, by Lemmas 2.8 and 2.2(4) the surface $\hat{F}_a$ is separating, hence u_i is parallel to u_j if and only if i and j have the same parity.

Without loss of generality we may assume that $val(u_1, \hat{\Gamma}_a^+) \leq 1$. Assume Γ_a is not kleinian. Then by Lemma 7.2 $\hat{\Gamma}_b^+$ has no boundary vertex of valence at most

7. THE CASE $n_a = 4$, $n_b \geq 4$ AND $\hat{\Gamma}_a^+$ HAS A SMALL COMPONENT

3, and by Lemma 6.4(3) it has no interior vertex of valence at most 7. Also, each vertex v_j of $\hat{\Gamma}_b^+$ has valence at least 3 because it is incident to at least three positive edges with label 1 at v_j, which by Lemma 2.3(3) must be mutually non-parallel. Therefore by Lemma 2.11 all vertices of $\hat{\Gamma}_b^+$ are boundary vertices of valence 4.

If $n_b > 4$ then by Lemma 2.3(3) the family of positive edges at u_1 contains at most $n_b/2 + 2 < n_b$ edges, so some v_j is incident to 4 positive edges with label 1 at v_j, which implies that v_j has at least 13 positive edges in four families, so one of the families contains 4 edges and hence Γ_a is kleinian by Lemma 6.4(1). Similarly if $\Delta > 4$ then Γ_a is kleinian.

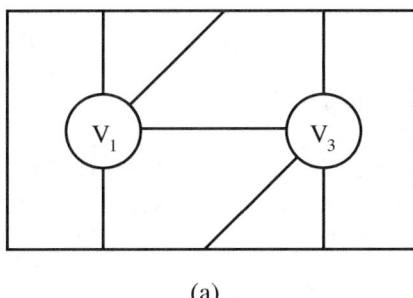 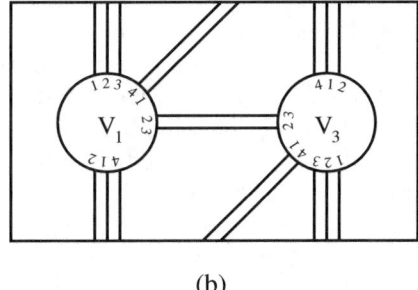

(a)   (b)

Figure 7.2

Now suppose $\Delta = n_b = 4$. Then $val(v_j, \hat{\Gamma}_b^+) = 4$ for all j implies that each component of $\hat{\Gamma}_b^+$ has two loops and two non-loop edges, as shown in Figure 7.2(a). By the parity rule a loop based at v_j has labels of different parity on its two endpoints, hence one sees that the number of positive edge endpoints of Γ_b at each v_j is even. By Lemma 6.4(4) we may assume that v_j has less than 12 positive edge endpoints, hence the above implies that each v_j is incident to exactly 10 positive edge endpoints. If some v_j is incident to only one loop in Γ_b then each of the non-loop family incident to v_j contains 4 edges and we are done. If some v_j is incident to two parallel loops in Γ_b then they form a Scharlemann bigon with label pair $\{1, 2\}$, say. Each of the two non-loop families contains three edges, hence the middle edge endpoint is a label of a Scharlemann bigon. Examining the labeling we see that all labels are Scharlemann bigon labels.

We now assume that each v_j is incident to three parallel loop edges. See Figure 7.2(b). The two outermost loops form a Scharlemann bigon with 1 as one of its labels. There are 6 adjacent negative edges at v_1, so three of then have labels 1 or 3 at v_1. By the same argument as in the last paragraph of the proof of Lemma 7.2 we may assume that the two Scharlemann bigons at v_1 and v_3 have the same label pair (12). The labeling of edge endpoints around v_1 and v_3 in a component of $\hat{\Gamma}_b^+$ is now as shown in Figure 7.2(b).

Because of the parity rule, the 4 non-loop edges cannot be divided into a family of 1 and another family of 3 edges, so they must form two pairs of parallel edges. From the labeling in Figure 7.2(b) one can see that they form two Scharlemann bigons with label pairs $\{2, 3\}$ and $\{4, 1\}$, respectively. The result now follows from Lemma 6.3. □

We now assume that $val(u_1, \hat{\Gamma}_a^+) = val(u_3, \hat{\Gamma}_a^+) = 2$. Then $\hat{\Gamma}_a^+$ contains either a cycle C containing both u_1, u_3, or it has two cycle components C, C' containing u_1, u_3, respectively.

LEMMA 7.4. *If $val(u_i, \hat{\Gamma}_a^+) = 2$ then i is a label of a Scharlemann bigon in Γ_b.*

PROOF. Let k be the number of interior vertices in $\hat{\Gamma}_b^+$. Let m be the number of edges in $\hat{\Gamma}_b^+$. We claim that $m \leq 2n_b + k$. Formally adding edges to $\hat{\Gamma}_b^+$ if necessary we may assume that any face A between two adjacent components of $\hat{\Gamma}_b^+$ is an annulus. The assumption $val(u_i, \hat{\Gamma}_a^+) = 2$ implies that no component of $\hat{\Gamma}_b^+$ is an isolated vertex. It is easy to see that if ∂A contains p vertices then we can add p edges to make each face on A a triangle. Therefore we can add at least $n_b - k$ edges to $\hat{\Gamma}_b^+$ to create a graph G on the torus $\hat{F}_b$ whose faces are all triangles. By an Euler characteristic argument we see that G has $3n_b$ edges, hence $\hat{\Gamma}_b^+$ has at most $3n_b - (n_b - k) = 2n_b + k$ edges, and the claim follows.

Now let m' be the number of negative edges of Γ_a incident to u_i. Note that if $m < m'$ then two negative edges at u_i are parallel in Γ_b and we are done. By Lemma 2.3(3) each positive family $\hat{e}$ in Γ_a contains at most $(n_b/2) + 2$ edges. Moreover, if $k > 0$ then some label does not appear on endpoints of edges in $\hat{e}$, so $\hat{e}$ has at most $n_b/2$ edges. Since u_i is incident to two families of positive edges, we have $m' \geq \Delta n_b - 2(n_b/2) \geq 3n_b$ if $k > 0$, and $m' \geq \Delta n_b - 2(n_b/2 + 2) \geq 3n_b - 4$ if $k = 0$. Since $m \leq 2n_b + k$, we have $m < m'$ (and hence i is a label of a Scharlemann bigon in Γ_b), unless $k = 0$, $\Delta = n_b = 4$ and $m = m' = 8$.

In this last case ($k = 0$, $\Delta = n_b = 4$ and $m = m' = 8$), all vertices of $\hat{\Gamma}_b^+$ are boundary vertices, hence by Lemma 6.8 there is no jumping label at u_i. On the other hand, since $m' = 8$, each positive family at u_i has 4 edges, so the two positive families cannot be adjacent by Lemma 6.6(2); hence there is a label j such that the two negative edges labeled j at u_i are separated by the two positive edges labeled j at u_i, so j is a jumping label at u_i, which is a contradiction. □

LEMMA 7.5. *If $val(u_1, \hat{\Gamma}_a^+) = val(u_3, \hat{\Gamma}_a^+) = 2$ then Γ_a is kleinian.*

PROOF. By Lemma 7.4 u_1, u_3 are labels of Scharlemann bigons. If some vertex, say u_4, is not a label of Scharlemann bigon then there must be (12)- and (23)-Scharlemann bigons in Γ_b. By Lemma 7.4 we have $val(u_4, \hat{\Gamma}_a^+) > 2$, so there is a loop edge e of $\hat{\Gamma}_a^+$ based at u_4. This is a contradiction to Lemma 7.1 (with labels permuted). □

PROPOSITION 7.6. *If $\hat{\Gamma}_a^+$ has a small component then (1) Γ_a is kleinian, and (2) $\hat{\Gamma}_a^+$ has at most 4 edges.*

PROOF. Let G be a small component of $\hat{\Gamma}_a^+$. If G contains only one vertex u_1 and two edges then it cuts the torus into a disk containing the other three vertices. It is easy to see that in this case there is a vertex of valence at most 2 in $\hat{\Gamma}_a^+$, which by Lemma 2.3(3) is incident to at most $2n_b$ edges, hence at least $2n_b$ negative edges. By Lemma 2.8 Γ_b has a Scharlemann cycle, so the surface $\hat{F}_a$ is separating. Therefore u_3 is parallel to u_1 and is antiparallel to u_2 and u_4. It follows that u_3 is incident to no positive edges, so by Lemma 7.3 Γ_a is kleinian. If G is not as above then either it contains a vertex of valence at most 1, or it is a cycle, in which case (1) follows from Lemmas 7.3 and 7.5.

Since Γ_a is kleinian, by Lemma 6.2(4) there is a free involution of Γ_a sending u_i to u_{i+1}, hence the number of edges ending at $\{u_2, u_4\}$ is the same as the number of edges ending at $\{u_1, u_3\}$, which is at most two in all cases discussed above. Hence (2) follows. $\square$

8. If $n_a = 4$, $n_b \geq 4$ and Γ_b is non-positive then $\hat{\Gamma}_a^+$ has no small component

Denote by X the union of $\hat{\Gamma}_b^+$ and all its disk faces.

LEMMA 8.1. *Suppose $n_a = 4$, $n_b \geq 4$, Γ_b is non-positive, and $\hat{\Gamma}_a^+$ has a small component. Then*

(1) each vertex of Γ_b is incident to at most 8 negative edges;

(2) if v_j is incident to more than 4 negative edges then j is a label of a Scharlemann bigon;

(3) if v_j is a boundary vertex of valence 3 in $\hat{\Gamma}_b^+$ then it is incident to either 6 or 8 negative edges, and j is a label of a Scharlemann bigon;

(4) $val(v_j, \hat{\Gamma}_b^+) \geq 3$ if v_j is a boundary vertex, and ≥ 2 otherwise;

(5) each component of X is either (a) a cyclic union of disks and (possibly) arcs, or (b) a cycle, or (c) an annulus.

PROOF. Since $\hat{\Gamma}_a^+$ has a small component, by Proposition 7.6 Γ_a is kleinian, and $\hat{\Gamma}_a^+$ has at most 4 edges.

(1) If v_j is incident to 9 negative edges then three of them are parallel on Γ_a because $\hat{\Gamma}_a^+$ has at most four edges, which contradicts Lemma 2.3(3).

(2) If v_j is incident to 5 negative edges then two of them form a Scharlemann bigon in Γ_a because $\hat{\Gamma}_a^+$ has only four edges by Proposition 7.6.

(3) Since Γ_a is kleinian, by Lemma 6.2(2) v_j is incident to an even number of negative edges. Each family of positive edges contains at most four edges, and by Lemma 6.6 two adjacent families contain at most 6 edges, hence the three positive families at v_j contain at most 10 edges. The result now follows from (1) and (2).

(4) By (1) v_j is incident to at least 8 positive edges, which are divided into at least two families, and if two then they cannot be adjacent by Lemma 6.6(1).

(5) If a component of X is contained in a disk then by Lemma 2.9 it would have either a boundary vertex of valence at most 2, which is impossible by (4), or six boundary vertices of valence 3, which is a contradiction because by (3) each such vertex is a label of Scharlemann bigon while by Lemma 2.3(4) $\hat{\Gamma}_b^+$ has at most two labels of Scharlemann bigons for each sign. Therefore no component of X is contained in a disk on the torus $\hat{F}_b$. Since Γ_b is not positive, this implies that each component of X is contained in an annulus but not a disk on $\hat{F}_b$.

If there is a sub-disk D of X such that $D \cap \overline{X - D}$ is a single point v then either $\hat{\Gamma}_b^+ \cap D$ contains a boundary vertex of valence 2 other than v, or 3 boundary vertices of valence 3 other than v, which again leads to a contradiction as above. $\square$

Let X_1 be a component of X, and let v_1 be a boundary vertex on the left cycle C_l of X_1, as defined in Section 4. Then there is another component X_2 of X such that the annulus A between C_l and the right cycle C'_r of X_2 has interior disjoint from $\hat{\Gamma}_b^+$. Denote by m_j the number of negative edges incident to v_j, and by $m' = m'_1$ the number of negative edges on A which are not incident to v_1.

LEMMA 8.2. *Suppose $n_a = 4$, $n_b \geq 4$, Γ_b is non-positive, and $\hat{\Gamma}_a^+$ has a small component. Let v_1 be a boundary vertex of X_1 with $m_1 > 4$. Then*

(1) v_1 is a label of a Scharlemann bigon;
(2) $m' = 0$ if $m_1 = 8$;
(3) $m' \leq 2$;
(4) C_l contains no other boundary vertices of valence at most 4.

PROOF. (1) Since $\hat{\Gamma}_a^+$ has only four edges, two of the negative edges at v_1 form a Scharlemann bigon on Γ_a.

(2) If $m_1 = 8$ then since $\hat{\Gamma}_a^+$ has only 4 edges, the 8 negative edges at v_1 form 4 Scharlemann cocycles, which must all go to the same vertex v_2 on C_r' because v_1 is a boundary vertex and the cocycles are essential loops. These cocycles separate C_l from C_r', hence all negative edges in A incident to a vertex of $C_l - v_1$ must have the other endpoint on v_2. On the other hand, by Lemma 8.1(1) v_2 is incident to at most 8 negative edges, and by the above all of them must connect v_2 to v_1. Hence $m' = 0$.

(3) By Proposition 7.6 Γ_a is kleinian, so by Lemma 6.2(2) m_1 is even; hence by (2) we may assume that $m_1 = 6$. Since $\hat{\Gamma}_a^+$ has only 4 edges, the 6 negative edges incident to v_1 contain at least 2 Scharlemann cocycles, which connect v_1 to some v_2 on C_r'. If v_2 is incident to 8 negative edges in A then as in (2) these edges form 4 Scharlemann cocycles, which must all connect to the same vertex v_1 and hence $m' = 0$. By Lemma 6.2(2) each family of parallel edges in Γ_b has an even number of edges, so v_2 cannot be incident to 7 negative edges in A. If v_2 is incident to 6 or less negative edges in A then by the above 4 of them connect to v_1, so there are at most 2 connecting to $C_l - v_1$, hence $m' \leq 2$.

(4) A boundary vertex on $C_l - v_1$ of valence at most 4 in $\hat{\Gamma}_b^+$ is incident to at most 12 positive edges by Lemma 6.6(2), and hence at least 4 negative edges, which must lie in A because it is a boundary vertex on C_l. This contradicts (3). □

LEMMA 8.3. *Suppose $n_a = 4$, $n_b \geq 4$, Γ_b is non-positive, and $\hat{\Gamma}_a^+$ has a small component. If a component X_1 of X contains a boundary vertex v_1 of valence 3, then X_1 is an annulus containing exactly two vertices, both of which are of valence 3 and are labels of Scharlemann bigons.*

PROOF. By Lemma 6.6(2) v_1 is incident to at least 6 negative edges. Consider the three possible types of X_1 in Lemma 8.1(5). It cannot be a cycle because it has a boundary vertex v_1. If X_1 is an annulus or a cyclic union of disks and arcs then by Lemma 8.2(4), $C_l - v_1$ has no boundary vertex of valence at most 4 in $\hat{\Gamma}_b^+$, which implies that there is a boundary vertex v_3 of valence 3 on the right circle C_r of X_1, hence for the same reason $C_r - v_3$ contains no boundary vertex of valence at most 4. By Lemma 8.2(3) there are at most 4 negative edges incident to $C_l \cup C_r - \{v_1, v_3\}$, so there is no (non-boundary) vertex of valence 2 on X_1. Thus either

(i) X_1 is an annulus containing only the two vertices v_1 and v_3; or

(ii) X_1 is an annulus containing exactly four vertices and the other two are boundary vertices of valence 5; or

(iii) X_1 is as in Figure 8.1 (a) or (b).

Case (i) gives the conclusion of the lemma because, as in the proof of Lemma 8.2, a boundary vertex of valence 3 must be a label of a Scharlemann bigon. We need to show that (ii) and (iii) are impossible.

8. THE CASE $n_a = 4$, $n_b \geq 4$ AND Γ_b POSITIVE

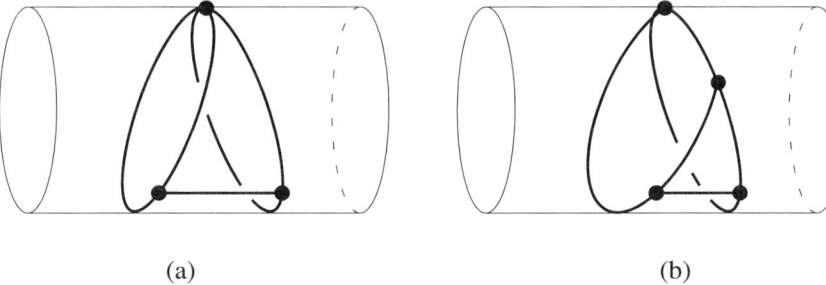

(a) (b)

Figure 8.1

In case (ii), let v_5, v_7 be the boundary vertices of valence 5 on X_1, with $v_5 \subset C_l$. Note that all faces of X_1 are triangles. If v_1 had 8 negative edges then by Lemma 8.2(2) v_5 would have no negative edge, which is impossible by Lemma 6.4(3). Therefore v_1 has exactly 6 negative edges and 10 positive edges, so by Lemma 6.6(1) the weights of the edges of X_1 incident to v_1 must be $(4, 2, 4)$. Similarly for v_3. Now the middle edge of X_1 at v_5 has weight 2 and the two boundary edges have weight 4, so again by Lemma 6.6(1) the weights around v_5 must be $(4, 2, 2, 2, 4)$. By Lemma 6.7(2) the edges of weight 2 must be non-Scharlemann. This is a contradiction because a triangle with a corner at v_5 bounded by two weight 2 edges has two non-Scharlemann bigons on its boundary while by Lemma 6.7(1) it has only one.

In case (iii), we assume X_1 is as in Figure 8.1(a). The other case is similar. Let v_1, v_3 be the vertices of valence 3 in the figure, and let v_5 be the other vertex. If $m_1 = 8$ then by Lemma 8.2(2) v_5 is incident to no negative edges on the side of C_l, and at most 2 negative edges on the side of C_r by Lemmas 8.1(3) and 8.2(3) applied to v_3. Therefore the four positive edges of $\hat{\Gamma}_b^+$ at v_5 are adjacent to each other, representing a total of at least 14 edges. It follows that there are two adjacent families of positive edges, each containing 4 edges, which is a contradiction to Lemma 6.6(1). If $m_1 = m_3 = 6$ then each of v_1 and v_3 is incident to 10 positive edges, so by Lemma 6.6(1) the weights of the edges of $\hat{\Gamma}_b^+$ at v_1 and v_3 are $(4, 2, 4)$, in which case v_5 again has two adjacent families of 4 positive edges each, contradicting Lemma 6.6(1). □

LEMMA 8.4. *Suppose $n_a = 4$, $n_b \geq 4$, Γ_b is non-positive, and $\hat{\Gamma}_a^+$ has a small component. If a component X_1 of X does not contain a boundary vertex of valence 3, then*

(1) X_1 is either a cycle or an annulus containing exactly two vertices; and
(2) all vertices of X_1 are labels of Scharlemann bigons.

PROOF. By Lemma 6.6(3) all interior vertices of X have valence at least 6. Since X_1 has no boundary vertex of valence 3, by Lemma 2.11 it is either a cycle, or an annulus with all interior vertices of valence 6, all boundary vertices of valence 4, and all faces triangles. The result is true when X_1 is a cycle because any vertex of valence 2 has more than 4 negative edges and hence is a label of Scharlemann cycle. Therefore we assume that X_1 is an annulus. If X_1 has an interior vertex v_1 then by Lemma 6.6(2) the weights of the edges of X_1 around v_1 must be $(4, 2, 4, 2, 2, 2)$ and any edge of weight 2 represents a non-Scharlemann bigon. Thus the triangle with a corner at v_1 bounded by two weight 2 edges has the property that it has at least two edges representing non-Scharlemann bigons, which is a contradiction to Lemma 6.7(1). Therefore X_1 has no interior vertex.

First assume that some vertex v_1 on the left cycle C_l of X_1 is incident to less than 12 positive edges, and hence more than 4 negative edges. Since all vertices of X_1 are of valence 4 in $\hat{\Gamma}_b^+$, by Lemma 8.2(3) C_l has no other vertices. In this case C_r also contains only a single vertex v_3. By Lemma 6.6(2) v_3 cannot have more than 12 positive edges, and if 12 then the weights around it are $(4,2,4,2)$. However this cannot happen because the first and the last numbers are the weights of the loop edge at v_3 and hence must be the same. Therefore v_3 has less than 12 positive edges. Now by Lemma 8.1(2) v_3 and v_1 are labels of Scharlemann bigons, and the result follows.

Now assume that each vertex of X_1 is incident to 12 positive edges. This implies that the weights of the edges of X_1 around each vertex are either $(4,2,4,2)$ or $(2,4,2,4)$, so any pair of adjacent edges have different weights. However, this is impossible because two of the three edges of a triangle face must have the same weight. □

LEMMA 8.5. *Suppose $n_a = 4$, $n_b \geq 4$, Γ_b is non-positive, and $\hat{\Gamma}_a^+$ has a small component. Then*

(1) $n_b = 4$.
(2) $val(u_i, \hat{\Gamma}_a^+) \geq 2$ for all i.
(3) Each component of $\hat{\Gamma}_a^+$ is a loop.

PROOF. (1) By Lemmas 8.3 and 8.4 each vertex of Γ_b is a label of a Scharlemann bigon, and by Lemma 2.3(4) there are at most 4 such labels. Hence $n_b = 4$.

(2) Suppose u_1 is incident to at most one edge of $\hat{\Gamma}_a^+$. By Lemmas 8.3 and 8.4 each component G of $\hat{\Gamma}_b^+$ consists of either (i) a cycle, or (ii) two loops and one non-loop edge, or (iii) two loops and two non-loop edges. Since there are at least 3 negative j-labels at u_1, there are at least 3 positive 1-labels at each v_j, hence (i) cannot happen. Moreover, since each v_j is a boundary vertex containing at least three positive 1-labels, it has more than 8 positive edges.

Suppose G is of type (ii). Then the label 1 appears three times on positive edge endpoints around each of the two vertices of G, hence it appears a total of 6 times among the three families of positive edges in G, so by Lemmas 2.4 and 6.2(3) there is a (12)-Scharlemann bigon among each of these families. Since a loop and a non-loop edge cannot be parallel in Γ_a (Lemma 2.3(5)), these represents at least four edges of $\hat{\Gamma}_a$ connecting u_1 to u_2, which cut the torus $\hat{F}_a$ into two disks. On the other hand, by the parity rule a loop at a vertex v of G must have labels of different parity on its two endpoints, so the total number of positive edges at v is at least 10, divided into three families, hence one of the families has four parallel edges, which contains a (34)-Scharlemann bigon, giving a pair of edges on Γ_a lying in the interior of the disks above which must therefore be parallel. This is a contradiction to the fact that a Scharlemann cocycle is essential (Lemma 2.2(5)).

Now suppose G is of type (iii). If some v_j is incident to 12 positive edges then by Lemma 6.6(1) the weights of the positive edges around v_j are $(4,2,4,2)$, which is impossible because the first and the last weights are for a loop and hence must be the same. Since v_j is incident to more than 8 positive edges, it is incident to exactly 10 positive edges, so the weights are $(2,4,2,2)$ or $(2,2,4,2)$ around each vertex. The two loops must be (12)-Scharlemann bigons in order for each vertex to have 3 edge endpoints labeled 1. This completely determines the labeling of the edge endpoints up to symmetry. Examining the labeling one can see that the family of 4 parallel edges form an extended Scharlemann cycle, which is a contradiction to Lemma 2.2(6).

(3) By Proposition 7.6 Γ_a is kleinian, hence the torus $\hat{F}_a$ is separating, so each edge of $\hat{\Gamma}_a^+$ has endpoints on vertices whose subscripts have the same parity. By (2) a small component of $\hat{\Gamma}_a^+$ must be a loop C. Let u_1 be a vertex of C. If C does not contain u_3 then the component of $\hat{\Gamma}_a^+$ containing u_3 must also be a loop because it contains no other vertices, and it cannot contain more than one edge as otherwise some component of $\hat{\Gamma}_a^+$ would lie in a disk and hence would have a vertex of valence at most 1, contradicting (2). Thus the graph G consisting of u_1, u_3 and all edges with endpoints on them is either one loop or two disjoint loops. By Lemma 6.2(4) there is a involution of $\hat{\Gamma}_a^+$ mapping u_1 and u_3 to u_2 and u_4 respectively, hence it maps G to $\hat{\Gamma}_a^+ - G$. Therefore the components in $\hat{\Gamma}_a^+ - G$ are also loops. □

PROPOSITION 8.6. *Suppose $n_a = 4$, $n_b \geq 4$, and Γ_b is non-positive. Then $\hat{\Gamma}_a^+$ has no small component.*

PROOF. Suppose to the contrary that $\hat{\Gamma}_a^+$ has a small component. By Lemma 8.5 we have $n_b = 4$ and each component of $\hat{\Gamma}_a^+$ is a loop. Thus each u_i is incident to at most 8 positive edges, so Γ_a has no more positive edges than negative edges.

By Lemmas 8.3 and 8.4, each component X_1 of X is either a circle or an annulus containing two vertices of $\hat{\Gamma}_b^+$. First assume that X_1 is a circle, so it is a small component of $\hat{\Gamma}_b^+$. Applying Proposition 7.6 and Lemma 8.5 with Γ_a and Γ_b reversed, we see that Γ_b is kleinian, and all components of $\hat{\Gamma}_b^+$ are also circles, hence Γ_b also has the property that it has no more positive edges than negative edges. Applying the parity rule we see that both graphs have the same number of positive edges and negative edges. In particular, each family of positive edges contains exactly 4 edges, which by Lemma 6.2(3) must consist of a (12)-Scharlemann bigon and a (34)-Scharlemann bigon. Dually it implies that all negative edges connect u_1 to u_2 or u_3 to u_4, so there are 4 families of negative edges, each containing exactly 4 edges. Now the two positive families at u_1 contain 4 edges each, and, whether separated by the two negative families or not, their endpoints at u_1 have the same label sequence. This contradicts Lemma 6.6(1).

We may now assume that X consists of two annular components X_1, X_2, each containing two vertices. Assume $v_1, v_3 \in X_1$. As in the last paragraph of the proof of Lemma 8.5(2), in this case each vertex of Γ_b is incident to 8 or 10 positive edges, therefore by Lemma 8.1(2) each vertex is a label of Scharlemann bigon, hence Γ_b is also kleinian.

If $val(v_1, \Gamma_b^+) = 10$ then v_1 is incident to at least 6 negative edges, which are divided into two families of parallel edges on the annulus bounded by the loops at v_1 and v_2. Since Γ_b is kleinian, each family contains an even number of edges, hence the number of edges in these two families are 4 and 2, respectively. Examining the labels at the endpoints of these edges, we see that two edges with the same label i at v_1 have different labels at v_2. On Γ_a this means that there are both loop and non-loop positive edges incident to u_i, which is a contradiction to the fact that $\hat{\Gamma}_a^+$ consists of cycles only.

We have shown that $val(v_j, \Gamma_b^+) = 8$ for all j. Thus there are 16 positive edges on Γ_a, so each of the two positive families incident to u_1 contains 4 edges. Since all vertices of Γ_b are boundary vertices, there is no jumping label on any vertex of Γ_b, hence by Lemma 6.8 there is no jumping label at u_1, so the two families of positive edges must be adjacent. This is a contradiction to Lemma 6.6(2). □

9. If Γ_b is non-positive and $n_a = 4$ then $n_b \leq 4$

Note that if Γ_a is positive then each vertex of Γ_b is a label of a Scharlemann bigon and hence $n_b \leq 4$ by Lemma 2.3(4). By Proposition 8.6 the statement in the title is true if $\hat{\Gamma}_a^+$ has a small component. Therefore we may assume that $\hat{\Gamma}_a^+$ consists of two large components G_1 and G_2, each of which must be one of the graphs of type (3), (9) or (10) in Figure 4.2. As before, denote by X_a the union of $\hat{\Gamma}_a^+$ and all its disk faces, and by X_i the components of X_a containing G_i, $i = 1, 2$. Denote $n = n_b$.

LEMMA 9.1. *Suppose that Γ_b is non-positive, $n_a = 4$ and $n > 4$. Then $\hat{\Gamma}_b^+$ contains no interior vertex.*

PROOF. Otherwise Γ_b has a vertex v_i which is incident to positive edges only. By Lemma 2.3(1) no three of these edges are parallel on Γ_a, so $\hat{\Gamma}_a$ contains at least $\Delta n_a/2 \geq 8$ negative edges, and hence at most $3n_a - 8 = 4$ positive edges by Lemma 2.5, so $\hat{\Gamma}_a^+$ has a small component, contradicting our assumption. □

LEMMA 9.2. *Suppose that Γ_b is non-positive, $n_a = 4$ and $n > 4$. Suppose X_a is a disjoint union of two annuli. Let G be the subgraph of Γ_b consisting of positive 1-edges and all vertices. Then G cannot have two triangle faces D_1, D_2 with an edge in common.*

PROOF. Since X_a is a disjoint union of two annuli, all negative edges of Γ_a incident to u_1 must have the other endpoint on the same vertex, say u_2, and vice versa. On G this means that every edge has label pair (12), and all positive edges with an endpoint labeled 1 or 2 are in G. Thus no edge in the interior of D_i has label 1 or 2 at any of its endpoints. Up to symmetry the labels on the boundary of the two triangles must be as shown in Figure 9.1. Since the labels 3 and 4 must appear between two label 1 at a vertex, one of the triangles, say D_1, must contain some (34) edges. Since all the vertices are parallel, one can see that the labels 3 and 4 appear at each corner of D_1, hence there are three edges inside of D_1. Since there is no trivial loop, they must form an extended Scharlemann cycle, contradicting Lemma 2.2(6). □

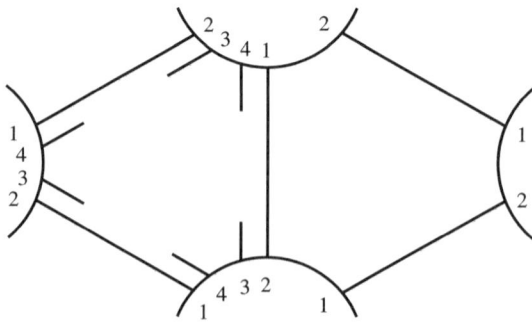

Figure 9.1

LEMMA 9.3. *Suppose that Γ_b is non-positive, $n_a = 4$ and $n > 4$. If X_a is a disjoint union of two annuli then Γ_a is kleinian.*

PROOF. By Lemma 6.3 it suffices to show that each vertex u_i of Γ_a is a label of a Scharlemann bigon. Let t_i be the number of negative edges at u_i. Since there are at most two families of negative edges incident to u_i, we have $t_i \leq 2n$ by Lemma 2.3(1). On the other hand, since u_i is a boundary vertex of valence at most 4 in $\hat{\Gamma}_a^+$, by Lemma 2.7 it is incident to at most $2(n+2) = 2n+4$ positive edges. If there are $2n+4$ then u_i is incident to at least four loops and the four outermost loops form an extended Scharlemann cycle, which is impossible by Lemma 2.2(6). Also by the parity rule the loops have labels of different parity at its two endpoints, hence the number of positive edges at u_i must be even. It follows that u_i is incident to either $2n$ or $2n+2$ positive edges, hence $t_i = 2n$ or $2n-2$.

Assume that u_1 is not a label of a Scharlemann bigon. Then the negative edges at u_1 are mutually non-parallel positive edges in Γ_b, hence $\hat{\Gamma}_b^+$ has at least $2n-2$ edges. Denote by G the subgraph of Γ_b consisting of positive 1-edges. Let Y be the union of G and its disk faces, and let k be the number of boundary vertices of G.

First assume $t_1 = 2n$. By Lemma 9.1 G has no interior vertex, and clearly it has no isolated vertex, so we can apply Lemma 2.10(1) to conclude that $k \geq t_1 - n = n$. Since G only has n vertices, we must have $k = n$, so all vertices of G are boundary vertices, and hence there is no cut vertex. In this case Y contains exactly n boundary edges, so it has at least one (actually n) interior edge e. Since equality holds for the above inequality, by Lemma 2.10(1) all faces of Y are triangles. Therefore e is the common edge of two adjacent triangle faces, which is a contradiction to Lemma 9.2. Therefore this case is impossible.

Now assume $t_1 = 2n - 2$. In this case the two outermost loops at u_1 form a Scharlemann bigon, so by Lemma 2.2(4) $\hat{F}_b$ is separating, hence two vertices of Γ_b are parallel if and only if they have the same parity. Therefore we can define G_1 (resp. G_2) to be the union of the components of G containing v_i with odd (resp. even) i. Similarly for Y_1 and Y_2. Then G_1 contains all the negative edges at u_1 with odd labels, and G_2 those with even labels. Therefore each G_i contains exactly $n-1$ edges.

The $2n+2$ positive edges at u_1 form at least $n+1$ negative edges in $\hat{\Gamma}_b$ because any family of Γ_b contains at most 2 such edges. Hence $\hat{\Gamma}_b$ contains at least $(n+1) + (2n-2) = 3n-1$ edges. Since a reduced graph on a torus contains at most $3n$ edges (Lemma 2.5), we may add at most one edge to make the faces of the graph all triangles. Hence $\hat{\Gamma}_b$ has at most one 4-gon and all other faces are triangles. In particular, one of the G_i, say G_1, has the property that all its faces are triangles.

Let V and E be the number of vertices and edges of G_1, and let E_b, V_b be the number of non-interior edges and boundary vertices, respectively. Note that $V - V_b$ is the number of cut vertices, and $E - E_b$ is the number of interior edges. We have shown that $V = n/2$ and $E = n - 1$.

By Lemma 2.10(1) we have $V_b \geq E - V = (n-1) - n/2$, hence G has $V - V_b \leq V - (E - V) = (n/2) - (n - 1 - n/2) = 1$ cut vertex. If there is no cut vertex then the number of non-interior edges is the same as the number of vertices V, i.e. $E_b = V$. If it has a cut vertex v then the equality $V_b = E - V$ holds, so by Lemma 2.10(1), v has exactly two corners not on disk faces, which implies v is incident to at most 4 non-interior edges, while every other vertex is incident to exactly two non-interior edges, hence $E_b \leq V + 1$. In either case we have

$E - E_b \geq E - (V+1) = (n-1) - (n/2+1) \geq 1$, so G has at least one interior edge e. Since all faces of Y_1 are triangles, e is incident to two triangle faces of G_1, which is a contradiction to Lemma 9.2. □

LEMMA 9.4. *Suppose that Γ_b is non-positive, $n_a = 4$ and $n > 4$. Then X_a is not a disjoint union of two annuli.*

PROOF. Assume to the contrary that X_a is a union of two annuli. Let t_i be the number of negative edges incident to u_i. As in the proof of Lemma 9.3, we have $t_i = 2n - 2$ or $2n$.

First assume that $t_1 = 2n - 2$. Let $\hat{e}_1, \hat{e}_2$ be the two families of edges in Γ_a connecting u_1 to u_2. Note that a (12)-Scharlemann bigon in Γ_b must have one edge in each of $\hat{e}_1$ and $\hat{e}_2$. By Lemma 9.3 Γ_a is kleinian, so all (12)-edges belong to Scharlemann bigons in Γ_b, hence each edge e_i in $\hat{e}_1$ is parallel in Γ_b to an edge e'_i in $\hat{e}_2$, and the label of e_i at u_1 is the same as that of e'_i at u_2. In particular, $\hat{e}_1$ and $\hat{e}_2$ have the same number of edges, hence each contains exactly $n - 1$ edges. Without loss of generality we may assume that the label n does not appear at the endpoints at u_1 of edges in $\hat{e}_1$. By the above, n does not appear at the endpoints at u_2 of edges of $\hat{e}_2$, hence the labels must be as in Figure 9.2. However, in this case the edge labeled 1 at u_1 has its other endpoint labeled n, which is a contradiction to the parity rule.

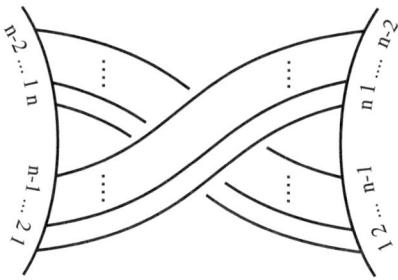

Figure 9.2

We now assume $t_1 = 2n$. Then the two families of negative edges from u_1 to u_2 have the same transition function φ. Since there is a (12)-Scharlemann bigon, $\varphi^2 = id$, so the length of each φ-cycle is 1 or 2. Since $n > 4$, it follows that the edges of $\hat{e}_1$ form at least 3 cycles on Γ_b, which is a contradiction to Lemma 2.14(2). □

LEMMA 9.5. *Suppose that Γ_b is non-positive, $n_a = 4$ and $n > 4$. If G_1 is of type (3), then G_2 is also of type (3), and the two loops of $\hat{\Gamma}_a^+$ do not separate the two vertices u_3, u_4 which are not on the loops.*

PROOF. Let u_3 be the vertex of G_1 which has valence 2. If G_2 is not of type (3), or if the two loops $\hat{e}_1 \cup \hat{e}_2$ of $\hat{\Gamma}_a^+$ separates u_3 and u_4 then all negative edges incident to u_3 have their other endpoint on the same vertex u_2 of G_2, and there are only two such families. Hence u_3 is incident to only four families of parallel edges, so by Lemmas 2.3(1) and 2.3(3) we must have $n = 4$, contradicting the assumption. □

LEMMA 9.6. *Suppose that Γ_b is non-positive, $n_a = 4$ and $n > 4$. Then $\hat{\Gamma}_a^+$ cannot be a union of two type (3) components.*

PROOF. Suppose that $\hat{\Gamma}_a^+$ is a union of two type (3) components, so $\hat{\Gamma}_a$ has 6 positive edges $\hat{e}_1, ..., \hat{e}_6$. By Lemma 9.5 the two loops do not separate the two vertices which are not on the loops, so the edges appear as in Figure 9.3. By Lemmas 2.3(1), 2.3(3) and 6.6(1) each vertex has valence at least 5, so there is one edge $\hat{e}_7$ from u_2 to u_3, two edges $\hat{e}_8, \hat{e}_9$ from u_1 to u_2, and one edge $\hat{e}_{10}$ from u_1 to u_4. There are one or two edges $\hat{e}_{11}$ and $\hat{e}_{12}$ connecting u_3 to u_4. See Figure 9.3.

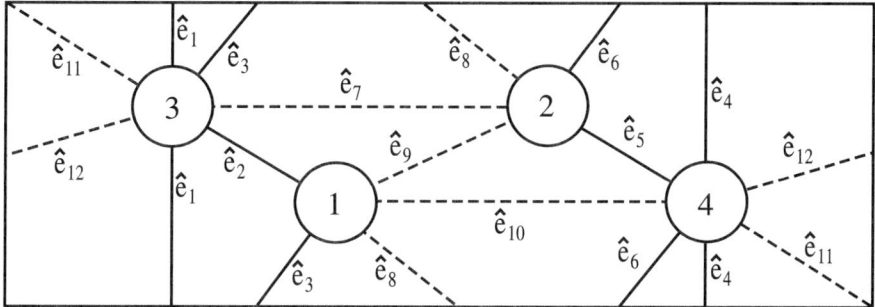

Figure 9.3

Denote by w_i the number of edges in $\hat{e}_i$. By Lemma 2.7(2) the two positive edges $\hat{e}_2, \hat{e}_3$ contain at most $n + 2$ edges, and the three negative families at u_1 contains at most $3n$ edges, hence $\Delta = 4$. Since each family contains at most n edges, we have the following inequalities.

(1) $$n \leq w_2 + w_3 \leq n + 2$$
(2) $$3n - 2 \leq w_8 + w_9 + w_{10} \leq 3n$$
(3) $$n - 2 \leq w_i \leq n \quad \text{for } i = 7, 8, 9, 10$$

Since u_1 has at least n adjacent positive edge endpoints and n adjacent negative edge endpoints, each vertex of Γ_b is incident to a positive edge and a negative edge, hence $\hat{\Gamma}_b^+$ has no isolated or interior vertex.

Claim 1. $\hat{\Gamma}_b^+$ has at least $2n - 2$ edges, hence Γ_a has at most two jumping labels.

Let $e \in \hat{e}_8$, and $e' \in \hat{e}_{10}$. Then e and e' are not parallel in Γ_b because if they were then they would form a Scharlemann bigon and hence have the same label pair on their endpoints, which is not the case because their label pairs are $\{1, 2\}$ and $\{1, 4\}$ respectively. Therefore the number of edges in $\hat{\Gamma}_b^+$ is at least

$$w_8 + w_{10} \geq 3n - 2 - w_9 \geq 2n - 2$$

Since $\hat{\Gamma}_b^+$ has no interior or isolated vertex, by Lemma 2.10(1) it has at most $2n - (2n - 2) = 2$ non-boundary vertices. Since these are the only vertices containing jumping labels, by Lemma 2.18 they are the only possible jumping labels of Γ_a.

Claim 2. $w_{11} + w_{12} \geq 2n - 2$.

Note that since $w_8, w_9 \leq n$, we have

$$\begin{aligned}
8n &= val(u_1, \Gamma_a) + val(u_2, \Gamma_a) \\
&= (w_2 + w_3 + w_{10}) + (w_5 + w_6 + w_7) + 2(w_8 + w_9) \\
&\leq (w_2 + w_3 + w_7) + (w_5 + w_6 + w_{10}) + 4n
\end{aligned}$$

Thus either $w_2 + w_3 + w_7 \geq 2n$ or $w_5 + w_6 + w_{10} \geq 2n$. Because of symmetry we may assume

(4) $$w_2 + w_3 + w_7 \geq 2n$$

Divide the edge endpoints on ∂u_3 into P_1, P_2, P_3, P_4, as shown in Figure 9.4. Denote by k_i the number of edge endpoints in P_i. A label that appears twice in one of the P_i will be called a *repeated label*. Note that if P_i contains a repeated label then $k_i > n$. Note also that a non-jumping label is a repeated label. Thus by Claim 1 there are at least $n - 2$ repeated labels among all the P_i. Since $k_1 = w_7 \leq n$, there is no repeated label in P_1.

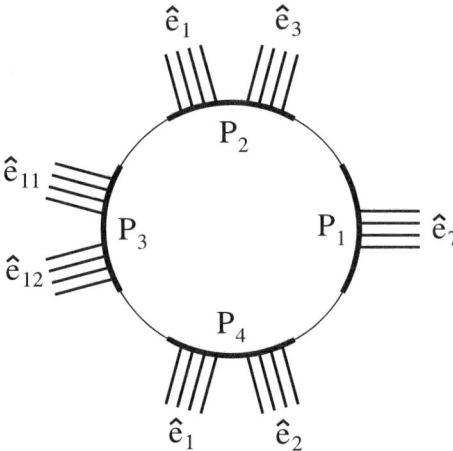

Figure 9.4

First assume that $k_3 > n$. If both $k_2, k_4 \leq n$ then all repeated labels are in P_3, so $k_3 \geq 2n - 2$ and we are done. Because of symmetry we may now assume to the contrary that $k_2 > n$. Thus one of $\hat{e}_1, \hat{e}_3$ contains more than $n/2$ edges, so by Lemmas 2.4 and 2.2(4) n is even, hence $n \geq 6$. Note that in this case $k_4 \leq n$, as otherwise we would have $4n = \sum k_i \geq (n-2) + 3(n+1) > 4n$, a contradiction. Thus P_2, P_3 contain all the repeated labels and there are at least $n - 2$ of them, so $k_2 + k_3 \geq 3n - 2$. Since $k_2 = w_1 + w_3 \geq n+1$ and $w_3 \leq (n/2) + 2$, we have $w_1 \geq (n/2) - 1$, so $w_3 - w_1 \leq 3$. By equation (4) above, we have

$$\begin{aligned} 4n &= val(u_3, \Gamma_a) = w_7 + k_2 + k_3 + (w_1 + w_2) \\ &= (k_2 + k_3) + (w_7 + w_3 + w_2) - (w_3 - w_1) \\ &\geq (3n - 2) + 2n - 3 = 5n - 5 > 4n \end{aligned}$$

This is a contradiction, which completes the proof for the case $k_3 > n$.

We now assume $k_3 \leq n$. Then each non-jumping label is a repeated label in either P_2 or P_4. By Lemma 2.7(2) we have $k_2, k_4 \leq n+2$, so each of P_2, P_4 contains at most 2 repeated labels, hence there are at most 4 repeated labels. On the other hand, we have shown that there are at least $n - 2$ repeated labels, so $n - 2 \leq 4$, i.e., $n \leq 6$. As above, P_2 or P_4 containing a repeated label implies that n is even, so $n = 6$; therefore there are at least 4 repeated labels. Hence there are exactly 4 repeated labels, so $k_2 = k_4 = n+2 = 8$. By Lemma 2.3(3) each of $\hat{e}_1, \hat{e}_2, \hat{e}_3$ contains exactly 4 edges.

By equation (3) above we have $n-2 \leq k_1 = w_7 \leq n$. If $k_1 = n-1$ or $n-2$ then the labels that do not appear in P_1 are repeated labels of both P_2 and P_4, so there are at most 3 distinct repeated labels, and hence at least $n-3 = 3$ jumping labels, contradicting Claim 1. If $k_1 = n$ then $k_1 + k_2 + k_4 = 3n+4$, so the endpoints of the 4 edges of $\hat{e}_1$ at P_2 have the same label sequence as those of $\hat{e}_1$ at P_4. Hence these edges form an extended Scharlemann cycle, which is impossible by Lemma 2.2(6).

Claim 3. Γ_a is kleinian.

Clearly each of u_1, u_2 is incident to more than $2n$ negative edges. Since $w_7 + w_8 + w_9 = 4n - (w_5 + w_6) \geq 3n - 2$, we have $w_7 \geq n - 2$. By Claim 2 $w_{11} + w_{12} \geq 2n - 2$, hence $w_{11} + w_{12} + w_7 \geq 3n - 4 > 2n$, so u_3 is also incident to more than $2n$ negative edges. Similarly for u_4. By Lemmas 9.1 and 2.10(2) $\hat{\Gamma}_b^+$ has at most $2n$ edges, so two of the negative edges at u_i are parallel on Γ_b, hence u_i is a label of Scharlemann bigon. Since this is true for all i, Γ_a is kleinian.

Claim 4. $w_{11} + w_{12} = 2n - 2$.

Since Γ_a is kleinian, by Lemma 6.2(4) there is a free involution η of Γ_a mapping u_1 to u_2 and u_3 to u_4. Thus η must map $\hat{e}_{10}$ to $\hat{e}_7$ and hence $w_7 = w_{10}$. We now have $w_2 + w_3 + w_7 = w_2 + w_3 + w_{10} \geq 4n - w_8 - w_9 \geq 2n$. Since $w_1 \geq 1$, we see that $w_{11} + w_{12} = 4n - (2w_1 + w_2 + w_3 + w_7) \leq 2n - 2$. The result now follows from Claim 2.

The involution η maps $\hat{e}_{11}$ to $\hat{e}_{12}$. Hence Claim 4 implies that $w_{11} = w_{12} = n-1$. Since η is label preserving, the label sequence of $\hat{e}_{11} \cup \hat{e}_{12}$ on ∂u_3 is the same as that on ∂u_4, so we may assume without loss of generality that the label sequences are as shown in Figure 9.2. One can see that in this case the transition function of $\hat{e}_{11}$ defined in Section 2 is transitive, which implies that all vertices of Γ_b are parallel, contradicting the assumption. This completes the proof of the lemma. □

PROPOSITION 9.7. *Suppose $n_a = 4$ and Γ_b is non-positive. Then $n_b \leq 4$.*

PROOF. Consider $\hat{\Gamma}_a^+$. If $\hat{\Gamma}_a^+$ has a small component then by Proposition 8.6 we have $n_b \leq 4$. If $\hat{\Gamma}_a^+$ has no small component then the component G containing u_1 must also contain u_3, and it is either of type (3), (9) or (10). The result follows from Lemma 9.4 if both components are of type (9) or (10), and from Lemmas 9.5 and 9.6 if at least one component is of type (3). □

10. The case $n_1 = n_2 = 4$ and Γ_1, Γ_2 non-positive

In this section we assume that $n_1 = n_2 = 4$ and Γ_a is non-positive for $a = 1, 2$. We will show that this case cannot happen. Denote by X_a the union of $\hat{\Gamma}_a^+$ and all its disk faces. By Theorem 8.6 $\hat{\Gamma}_a^+$ has no small component, so each component of $\hat{\Gamma}_a^+$ is of type (3), (9) or (10) in Figure 4.2.

LEMMA 10.1. *Suppose $n_1 = n_2 = 4$, and both Γ_1 and Γ_2 are non-positive. Then at least one component of $\hat{\Gamma}_1^+$ or $\hat{\Gamma}_2^+$ is of type (3).*

PROOF. Suppose to the contrary that all components of $\hat{\Gamma}_1^+$ and $\hat{\Gamma}_2^+$ are of type (9) or (10). Then each component of X_a is an annulus, hence any vertex u_i of Γ_a is incident to at most 2 families of negative edges. By Lemma 2.3(1) each negative family contains at most 4 edges, so u_i is incident to at most 8 negative edges, and hence the number of negative edges is no more than the number of positive edges in Γ_a. Since this is true for $a = 1, 2$ and since a positive edge in one graph is a negative edge in the other, the numbers of positive and negative edges of Γ_a must be the same, hence each vertex must be incident to exactly 8 positive and 8 negative edges, so each negative family contains exactly 4 edges. Since Γ_b contains a loop, one of the negative families in Γ_a contains a co-loop and hence is a set of 4 parallel co-loops, which is a contradiction to the 3-Cycle Lemma 2.14(2). □

LEMMA 10.2. *Suppose $n_1 = n_2 = 4$, and both Γ_1 and Γ_1 are non-positive. If both components of $\hat\Gamma_a^+$ are of type (3), and no component of $\hat\Gamma_b^+$ is of type (3), then Γ_a is kleinian.*

PROOF. Note that in this case all vertices of Γ_b are boundary vertices. Let u_1, u_2 be the vertices of valence 2 in $\hat\Gamma_a^+$. By Lemma 6.6(1), u_1 is incident to fewer than 8 positive edges, hence there are three negative edges in Γ_a incident to u_1 having the same label j at u_1. On Γ_b this implies that the vertex v_j is incident to at least three positive edges with label 1 at v_j; since v_j is a boundary vertex, it is incident to at least 9 positive edges. Since a loop at v_j must have labels of different parity on its two endpoints, we see that $val(v_j, \Gamma_b^+)$ is even. If $val(v_j, \Gamma_b^+) = 12$ then by Lemma 6.4(4) Γ_a is kleinian. Hence we may assume that $val(v_j, \Gamma_b^+) = 10$. By Lemma 6.4(1) we may assume that each family of positive edges at v_j has at most 3 edges. This implies that v_j is incident to 2 or 3 loops. Examining the labels we see that the two outermost loops form a Scharlemann bigon, with 1 as a label. For the same reason, 2 is a label of a Scharlemann bigon in Γ_b. If there is no (12)-Scharlemann bigon then there must be (14)- and (23)-Scharlemann bigon, so Γ_a is kleinian and we are done. Therefore we may assume that Γ_b contains a (12)-Scharlemann bigon.

The (12)-Scharlemann bigon and $\hat\Gamma_a^+$ cuts $\hat F_a$ into faces. There is now only one edge class in these faces which connects u_1 to u_4, hence by Lemma 2.2(5) Γ_b contains no (14)-Scharlemann bigon. Similarly there is no (23)-Scharlemann bigon. It follows that all Scharlemann bigons of Γ_b have label pair (12). In particular, the two outermost loops at v_j must form a (12)-Scharlemann bigon.

We have shown above that the vertex v_j has 2 or 3 loops. If it has 2 loops then the weights of the positive families at v_j are $(2, 3, 3, 2)$, and the middle label of a family of 3 is a label of Scharlemann bigon, which implies that both 3 and 4 are labels of Scharlemann bigons, which is a contradiction. Hence v_j has exactly 3 loops e_1, e_2, e_3, and 4 non-loop edges divided to 2 non-Scharlemann bigons $e_4 \cup e_5$ and $e_6 \cup e_7$.

As shown above, $e_1 \cup e_2$ is a (12)-Scharlemann bigon, so up to symmetry we may assume that the edges e_4, e_5 have labels 4 and 1 at v_j. Since these edges do not form a Scharlemann bigon, the labels at their other endpoints must be 3 and 2 respectively, so e_5 is a (12)-edge, which must be parallel on Γ_a to one of the two (12)-loops e_1, e_2 because Γ_a has only two families connecting u_1 to u_2. This is a contradiction because by Lemma 2.3(5) a loop and a non-loop edge cannot be parallel on Γ_a. □

LEMMA 10.3. *Suppose $n_1 = n_2 = 4$, and both Γ_1 and Γ_2 are non-positive. Then all components of $\hat\Gamma_1^+$ and $\hat\Gamma_2^+$ are of type (3), and $\hat\Gamma_1$ and $\hat\Gamma_2$ are subgraphs of the graph shown in Figure 10.1.*

10. THE CASE $n_1 = n_2 = 4$ AND Γ_1, Γ_2 NON-POSITIVE

PROOF. By Lemma 10.1 we may assume that $\hat{\Gamma}_a^+$ has a component C of type (3). Let u_1 be the valence 2 vertex in C. If u_1 is incident to at most 4 families of parallel edges then each family contains exactly 4 edges, but since the two positive families are adjacent, this would be a contradiction to Lemma 6.6(1). Therefore u_1 is incident to at least 5 families of edges. Note that if the other component C' of $\hat{\Gamma}_a^+$ is not of type (3), or if it is of type (3) but the loop of C' separates u_1 from the valence 2 vertex u_2 of C' then u_1 would have only two families of negative edges, which is a contradiction. It follows that C' is also of type (3), so the graph $\hat{\Gamma}_a$ is a subgraph of that in Figure 10.1.

For the same reason if some component of $\hat{\Gamma}_b^+$ is of type (3) then $\hat{\Gamma}_b$ is a subgraph of that in Figure 10.1 and we are done. Therefore we may assume that no component of $\hat{\Gamma}_b^+$ is of type (3). By Lemma 10.2 Γ_a is kleinian.

Consider the edges $\hat{e}_1, ..., \hat{e}_6$ of $\hat{\Gamma}_a$ incident to u_3. See Figure 10.1. Let p_i be the weight of $\hat{e}_i$. By Lemmas 6.2(2) and 2.3(1), p_i is even and at most 4. Note that p_1, p_2, p_3 are non-zero, so by Lemma 6.6(1) we know that $p_1 + p_2$ and $p_1 + p_3$ are between 4 and 6. If $p_4 = 0$ then u_2 would have valence 4 in $\hat{\Gamma}_a$, which would lead to a contradiction as above. Hence $p_4 > 0$. If $p_5 + p_6 = 0$ then $val(u_3, \hat{\Gamma}_a^+) = 5$, so either $p_1 = 4$ or $p_2 = p_3 = p_4 = 4$; either case contradicts Lemma 6.6(1). Therefore $p_5 + p_6 \geq 2$.

One can now check, from the labeling around the boundary of u_3, that if $p_4 = 2$ then both labels at the edge endpoints of $\hat{e}_4$ at u_3 are jumping labels at u_3, and if $p_4 = 4$ then all labels at the endpoints of $\hat{e}_5 \cup \hat{e}_6$ at u_3 are jumping labels. This is a contradiction because all vertices of $\hat{\Gamma}_b^+$ are boundary vertices and hence by Lemma 6.8 u_3 should have no jumping label. □

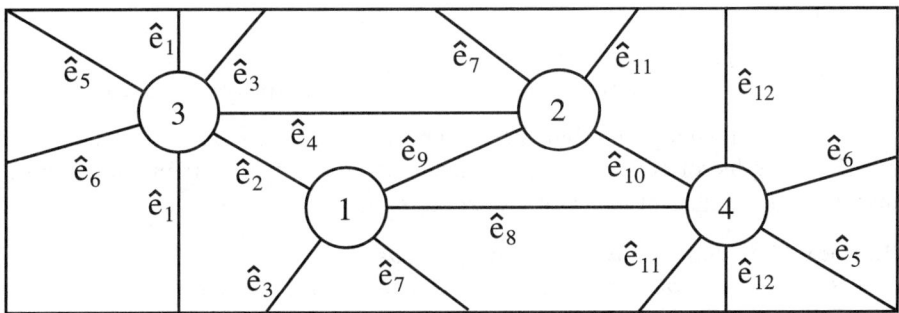

Figure 10.1

By Lemma 10.3 we may assume that both $\hat{\Gamma}_1, \hat{\Gamma}_2$ are subgraphs of that in Figure 10.1. Label the edges of $\hat{\Gamma}_a$ as in Figure 10.1, and let p_i be the weight of $\hat{e}_i$. Label $\hat{\Gamma}_b$ similarly using $\hat{e}_i'$.

LEMMA 10.4. $\Delta = 4$, $p_2 + p_3 = 6$, $p_8 = p_4 = 2$, and $p_7 = p_9 = 4$. Moreover, edges in $\hat{e}_4 \cup \hat{e}_8$ are co-loops, while those in $\hat{e}_7 \cup \hat{e}_9$ are not.

PROOF. If $p_2 + p_3 > 6$ then one of the $\hat{e}_2, \hat{e}_3$ contains 4 edges, so by Lemma 6.4(1) Γ_b is kleinian. By Lemma 6.2(2) the p_i are even, so $p_2 + p_3 > 6$ implies that $p_2 = p_3 = 4$, which is a contradiction to Lemma 6.6(2). Therefore we have

$p_2 + p_3 \le 6$. Since each $p_i \le 4$, by counting edges at u_1 we have $\Delta = 4$, and $p_7 + p_8 + p_9 \ge 10$.

Recall that if a pair of negative edges of Γ_a incident to u_i are parallel in Γ_b then they form a Scharlemann bigon in Γ_b, hence must have the same label pair in Γ_b, so they are incident to the same pair of vertices in Γ_a. Therefore no edge in $\hat{e}_7$ or $\hat{e}_9$ is parallel in Γ_b to an edge in $\hat{e}_8$. Since $\hat{\Gamma}_b^+$ has at most 6 edges, we have $\max\{p_7, p_9\} + p_8 \le 6$. Since $p_i \le 4$, this gives $p_7 + p_8 + p_9 \le 10$, hence $p_2 + p_3 = \Delta n_b - (p_7 + p_8 + p_9) \ge 6$. Together with the inequalities above, we have $p_2 + p_3 = 6$, $p_7 + p_8 + p_9 = 10$. Since $\max\{p_7, p_9\} + p_8 \le 6$ and $p_i \le 4$, this holds only if $p_8 = 2$ and $p_7 = p_9 = 4$. Similarly $p_4 = 2$.

We have shown that edges in $\hat{e}_7 \cup \hat{e}_8$ belong to distinct families in $\hat{\Gamma}_b^+$. Since $\hat{\Gamma}_b^+$ has at most 6 edges, $\hat{e}_7 \cup \hat{e}_8$ represent all edges in $\hat{\Gamma}_b^+$. If some edge in $\hat{e}_i$ is a co-loop then all of them are. Therefore the edges in $\hat{e}_7$ cannot be co-loops because $\hat{\Gamma}_b^+$ has only two loops. It follows that the edges in $\hat{e}_8$ must be co-loops. Similarly, edges of $\hat{e}_4$ are co-loops, and those in $\hat{e}_9$ are not. □

PROPOSITION 10.5. *Suppose both Γ_1 and Γ_2 are non-positive, and $n_a = 4$. Then $n_b < 4$.*

PROOF. By Proposition 9.7 we have $n_b \le 4$. Assume to the contrary that $n_b = 4$. Since the two edges in $\hat{e}_8$ are co-loops, they have labels $3, 4$ at u_1. Consider the three negative edges e_7, e_8, e_9 such that $e_i \in \hat{e}_i$, and they all have label 3 at u_1. In Γ_b these are 1-edges at v_3. Since e_8 is a loop, it belongs to $\hat{e}'_1$. The other two edges are non-loop positive edges on Γ_b, so they belong to $\hat{e}'_2 \cup \hat{e}'_3$. Applying Lemma 10.4 to Γ_b, we see that $\hat{e}'_2 \cup \hat{e}'_3 \cup \hat{e}'_4$ contains 8 edges, so the two edges e_7, e_9 are adjacent among the four edges labeled 1 at v_3 in Γ_b. Since they are not adjacent among the four edges labeled 3 at u_1 in Γ_a, this is a contradiction to the Jumping Lemma 2.18. □

11. The case $n_a = 4$, and Γ_b positive

In this section we assume that $n_a = 4$ and Γ_b is positive. We will determine all the possible graphs for this case. Recall from Lemma 6.4(2) that in this case Γ_a is kleinian, so the weights of edges of $\hat{\Gamma}_b$ are all even.

LEMMA 11.1. *Suppose $n_a = 4$ and Γ_b is positive.*

(1) Two families of 4 parallel edges with the same label sequence at a given vertex v_j of Γ_b connect v_j to the same vertex v_k.

(2) There are at most three families of 4 parallel edges with the same label sequence at any vertex v_j, and if $n_b > 2$ then there are at most two such.

(3) if $\Delta = 4$ then $val(v_j, \hat{\Gamma}_b) \ge 5$ for all j;

(4) if $\Delta = 5$ then $val(v_j, \hat{\Gamma}_b) \ge 6$ for all j;

(5) two weight 4 edges $\hat{e}_1, \hat{e}_2$ of $\hat{\Gamma}_b$ adjacent at a vertex v_j form an essential loop on $\hat{F}_b$.

PROOF. (1) If there are two families of 4 parallel edges with the same label sequence $1, 2, 3, 4$ at v_j then by Lemma 6.5 the initial edges e_1, e'_1 of the two families are parallel in Γ_a, with the same label j at the vertex u_1, hence the other endpoints of e_1 and e'_1 must also have the same label k, which implies that in Γ_b the two families have the same endpoints.

(2) If there were four then the leading edges of the (12) Scharlemann bigons in these families are parallel in Γ_a, so they belong to a family of at least $3n_b + 1$ parallel edges connecting u_1 to u_2. Since $\hat{\Gamma}_b$ has at most $3n_b$ edges by Lemma 2.5, two of these edges would be parallel on both graphs, which is a contradiction to Lemma 2.2(2).

If $n_b > 2$ and there are three families of 4 parallel edges with the same label sequence at v_j then as above there would be a family of $2n_b + 1$ parallel edges in Γ_a, which contradicts Lemma 2.22(3).

(3) and (4) follow immediately from (2).

(5) By (1) these two edges have their other endpoints at the same vertex v_k, hence form a loop $C = \hat{e}_1 \cup \hat{e}_2$ on $\hat{F}_b$. If C is not essential then we can choose C to be an innermost such cycle. C bounds a disk D on $\hat{F}_b$, which must contain some vertex because $\hat{\Gamma}_b$ is reduced. If some vertex in the interior of D has valence 5 then it is incident to two adjacent weight 4 edges, which would form another inessential loop, contradicting the choice of C. Hence all vertices in the interior of D have valence at least 6. By Lemma 2.9 in this case there should be at least three vertices on ∂D, which is a contradiction. □

$\hat{F}_a$ separates $M(r_a) = M \cup V_a$ into the *black* and *white* sides X_B and X_W. Since Γ_b is kleinian (Lemma 6.4(2)), the black side X_B is a twisted I-bundle over the Klein bottle. A face of Γ_b is white if it lies in the white region X_W, otherwise it is black. In the next two lemmas we assume that Γ_b contains a white bigon and a white 3-gon as in Figure 11.1.

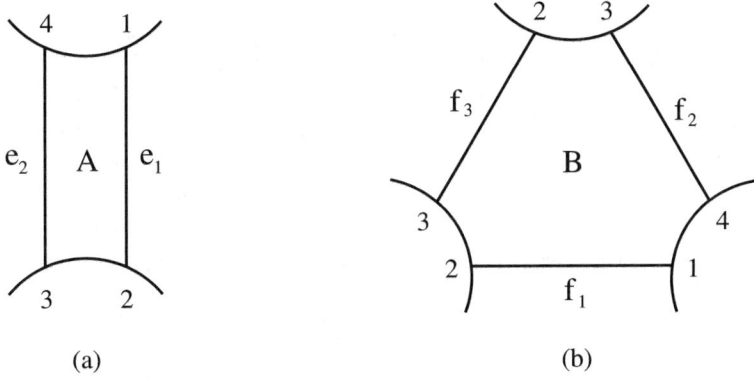

Figure 11.1

LEMMA 11.2. *Suppose $n_a = 4$ and Γ_b is positive. Suppose Γ_b contains a white bigon A and a white 3-gon B as in Figure 11.1. Then up to homeomorphism of $\hat{F}_a$, the edges of A and B appear on $\hat{F}_a$ as shown in Figure 11.2.*

PROOF. Let V_{23} and V_{41} be the components of $V_a \cap X_W$ that run between u_2, u_3 and u_4, u_1, respectively. Let Y be a regular neighborhood of $\hat{F}_a \cup V_{23} \cup V_{41} \cup A \cup B$. Then $\partial Y = \hat{F}_a \cup T$, where T is a torus in M, and hence either $X_W = Y$, or X_W is the union of Y and a solid torus along T.

Take a regular neighborhood D of $e_1 \cup e_2 \cup f_3$ as "base point" for $\pi_1(\hat{F}_a) \cong \mathbb{Z} \times \mathbb{Z}$ and for $\pi_1(X_W)$. (See Figure 11.2). The cores of the 1-handles V_{23} and

V_{41} represents elements x, y respectively of $\pi_1(X_W)$, and $\pi_1(X_W)$ is generated by $\Pi = \pi_1(\hat{F}_a)$ together with x and y. Note that since $\hat{F}_a$ is essential in $M(r_a)$, Π is a proper subgroup of $\pi_1(X_W)$.

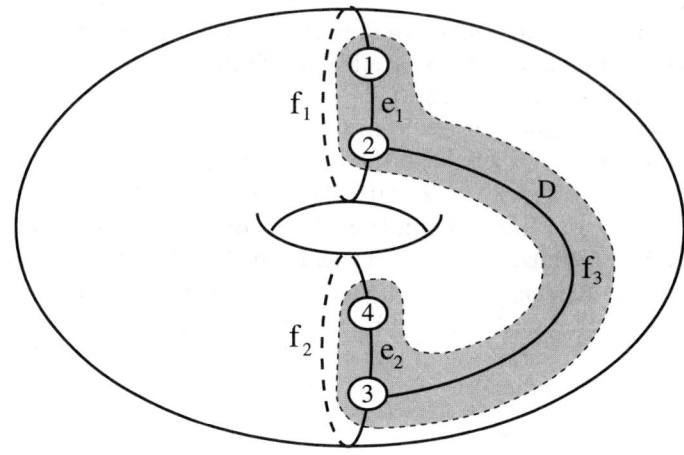

Figure 11.2

The bigon A gives the relation $xy = 1$, so the 23-corners and the 41-corner of ∂B represent x and x^{-1} respectively. The 32-edge on ∂B lies in D and hence represents 1. If one of the other two edges f_1, f_2 of B represents $1 \in \pi_1(\hat{F}_a)$, then B gives a relation of the form $x = \gamma$ for some $\gamma \in \pi_1(\hat{F}_a)$, hence $\pi_1(X_W) = \Pi$, a contradiction. Therefore the edge f_1 is as shown in Figure 11.2, and represents a nontrivial element $\alpha \in \pi_1(\hat{F}_a)$, when oriented from u_1 to u_2. Similarly, the edge f_2 represents a non-trivial element γ, say, of $\pi_1(\hat{F}_a)$, when oriented from u_3 to u_4. Since $e_1 \cup f_1$ and $e_2 \cup f_2$ are disjoint, we must have $\gamma = \alpha$ or α^{-1}. The union $V_{23} \cup V_{41} \cup N(A)$ is a single 1-handle attached to $\hat{F}_a$, and ∂B is a simple closed curve on $\hat{F}_a \cup \partial H$. One can see that $\alpha x^2 \alpha^{-1} x^{-1}$ cannot be realized by a simple closed curve. It follows that $\gamma = \alpha$, so f_2 appears on $\hat{F}_a$ as in Figure 11.2. □

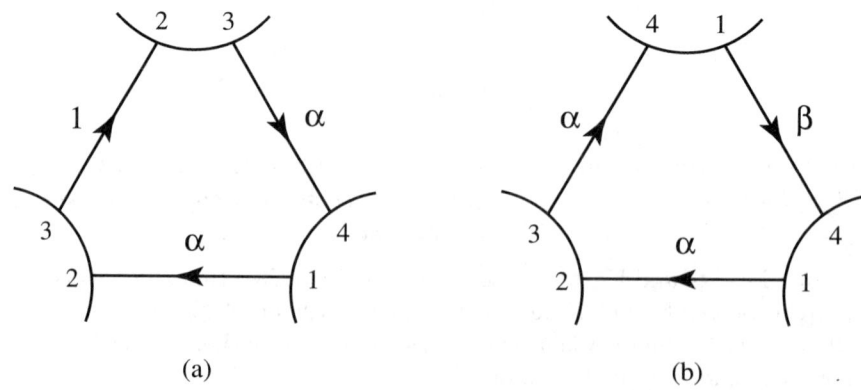

Figure 11.3

11. THE CASE $n_a = 4$, AND Γ_b POSITIVE

If we orient an edge e of Γ_b then the corresponding oriented edge of Γ_a represents an element γ of $\pi_1(\hat{F}_a)$, and we will label the edge e with γ. Thus the edge labels of the bigon A are both 1, and the edge labels of the 3-gon B are as in Figure 11.3(a). Note that any 12-edge, oriented from 1 to 2, or any 34-edge, oriented from 3 to 4, has label 1 or α. Also $\pi_1(\hat{F}_a)$ has a basis $\{\alpha, \beta\}$, where β is represented by an arc joining u_1 and u_4, disjoint from ∂A and ∂B, as shown in Figure 11.4.

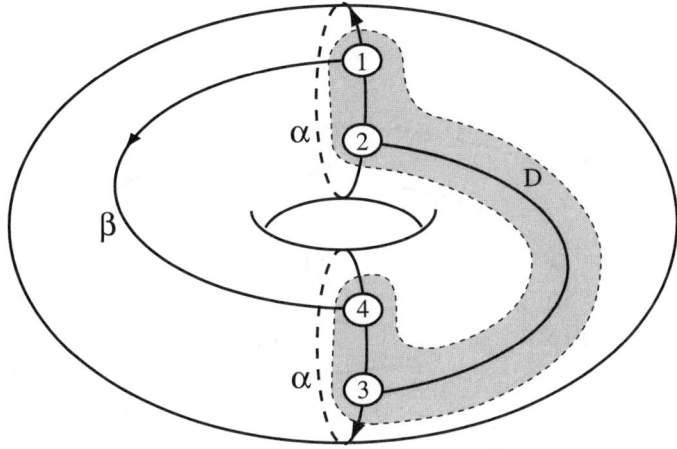

Figure 11.4

LEMMA 11.3. *Suppose $n_a = 4$ and Γ_b is positive. Suppose Γ_b contains a white bigon A and a white 3-gon B as in Figure 11.1.*

(1) All edges of white bigons have label 1.

(2) β can be chosen so that any 3-gon must have edge labels as shown in Figure 11.3(a) or (b).

PROOF. (1) Since the edges $e_1 \cup f_1$ and $e_2 \cup f_2$ on ∂A and ∂B form two parallel essential circles on $\hat{F}_a$, any 12-edge of Γ_b must be parallel to either e_1 or f_1, and any 34-edge of Γ_b must be parallel to either e_2 or f_2.

Let A' be another white bigon. Applying Lemma 11.2 to A' and B gives that the 12-edge of A' and the 12-edge f_1 of B are not parallel, and similarly the 34-edge of A' and the 34-edge f_2 of B are not parallel. Hence both edges of A' are labeled 1 since they must be parallel to e_1 and e_2, respectively.

(2) Since Γ_b has no extended Scharlemann cycle, each triangle face B' has either one or two 23-corners. If B' has two 23-corners then applying Lemma 11.2 to A and B' shows that the 12-edge of B' is not parallel to e_1, so it must be parallel to f_1 and hence is labeled α, as in Figure 11.3(a). Similarly the 34-edge of B' is also labeled α. If the 32-edge of B' is labeled γ then B and B' together give the relation $\gamma = 1$, so $\gamma = 1$, as in Figure 11.3(a).

If B' has only one 23-corner, let f_1', f_2', f_3' be the 12-, 34- and 14-edges of B', respectively. Applying Lemma 11.2 to A and B' gives that f_1' is not parallel to e_1, so by the above it must be parallel to f_1 and hence is labeled α. Similarly f_2' is parallel to f_2 and is also labeled α. The two loops $e_1 \cup f_1$ and $e_2 \cup f_2$ cut $\hat{F}_a$ into two annuli A_L and A_R, where A_R contains f_3; see Figure 11.4. If the 14-edge f_3' of B' lies in A_R then it is labeled α, and B' gives the relation $x^{-1}\alpha x \alpha x^{-1}\alpha = 1$. It is easy to see that, together with the relation $x^2\alpha x^{-1}\alpha = 1$ coming from B, this implies $\alpha = x^2$, so $x^5 = 1$, and hence $\alpha^5 = 1$, which is a contradiction to the fact that α^5 is a nontrivial element in $\pi_1(\hat{F}_a)$ and hence is nontrivial in $\pi_1(X_W)$. Therefore f'_3 lies in A_L, as shown in Figure 11.4. Let β be the corresponding element of $\pi_1(\hat{F}_a)$; then the edge labels of B' are as in Figure 11.3(b). If B'' is any other 3-gon with one 23-corner then the argument above in the case of 3-gons with two 23-corners, using A and the present B', shows that the edge labels of B'' are also as shown in Figure 11.3(b). □

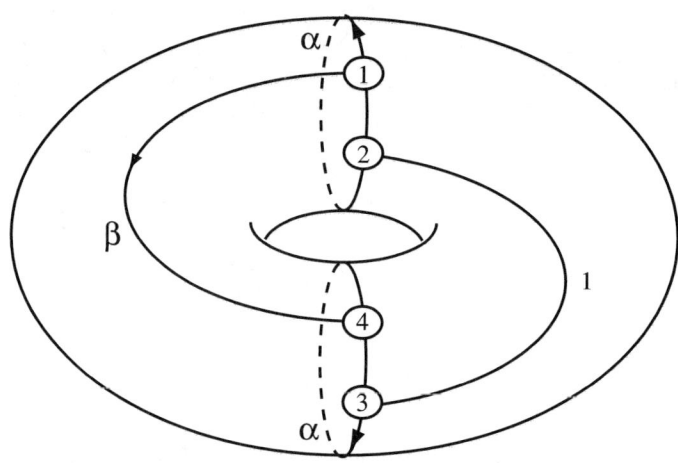

Figure 11.5

COROLLARY 11.4. *Suppose $n_a = 4$ and Γ_b is positive.*

(1) Let G be the subgraph of Γ_a consisting of edges of white bigons and white 3-gons on Γ_b. Then the reduced graph $\hat{G}$ is a subgraph of that in Figure 11.5.

(2) If Γ_b contains a white bigon then it cannot contain a (black) Scharlemann bigon which is flanked on each side by a (white) 3-gon.

(3) If Γ_b contains a white bigon then it cannot contain three 3-gons occurring as consecutive white faces at a vertex.

PROOF. (1) This follows immediately from Lemma 11.3.

(2) The edges of a (black) Scharlemann bigon are either (12)- or (34)-edges, so by Lemma 11.3(2) both edges of the Scharlemann bigon are labeled α and hence are parallel on Γ_a, contradicting Lemma 2.2(2).

(3) By (2) the two black bigons between the white 3-gons are (12, 34)-bigons, as shown in Figure 11.6. But then the middle 3-gon would be a white Scharlemann cycle, contradicting Lemma 6.2(3). □

11. THE CASE $n_a = 4$, AND Γ_b POSITIVE

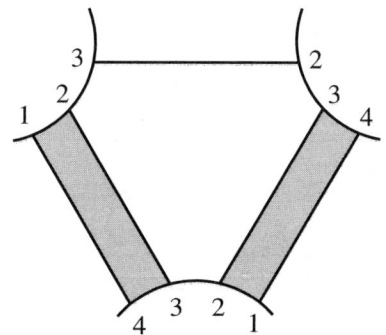

Figure 11.6

Let G be a reduced graph on a torus T with disk faces. One can endow T with a singular Euclidean structure by letting each edge have length 1 and each n-gon face a regular Euclidean n-gon. The cone angle $\theta(v)$ at a vertex v of G is the sum of the angles of the corners incident to v. Such a structure is *hyperbolic* if $\theta(v) \geq 2\pi$ for all v, and $\theta(v) > 2\pi$ for some v. The following lemma says that no singular Euclidean structure on T is hyperbolic.

LEMMA 11.5. *Let G and $\theta(v)$ be defined as above. Then either $\theta(v) < 2\pi$ for some v, or $\theta(v) = 2\pi$ for all v.*

PROOF. Denote by V, E, F the numbers of vertices, edges and faces of G, respectively. Let $\theta(c)$ be the angle at a corner c of the graph. If σ is a face of G, denote by $|\sigma|$ the number of edges of σ. Since σ is a regular $|\sigma|$-gon, for each corner $c \in \sigma$ we have $\theta(c) = \pi(1 - 2/|\sigma|)$. In the following, the first sum is over all vertices v of G, and the second is over all corners c. Grouping corners by faces σ, we get

$$\sum_v \theta(v) \quad = \sum_c \theta(c) = \sum_\sigma \sum_{c \in \sigma} \theta(c) = \sum_\sigma \sum_{c \in \sigma} \pi(1 - \tfrac{2}{|\sigma|})$$

$$= \pi(\sum_\sigma |\sigma| - \sum_\sigma \sum_{c \in \sigma} \tfrac{2}{|\sigma|}) = \pi(2E - 2F) = 2\pi(E - F) = 2\pi V.$$

Therefore $\sum_v (2\pi - \theta(v)) = 0$, and the result follows. □

LEMMA 11.6. *Let G be a reduced graph on a torus T such that $val(v) \geq 5$ for all v. Then either*

(1) there exists a vertex of valence 5 with at least four 3-gons incident; or

(2) there exists a vertex of valence 6 and all vertices of valence 6 have all incident faces 3 gons; or

(3) all faces of G are 3-gons or 4-gons, and every vertex has valence 5 and has exactly three 3-gons incident.

PROOF. We have $\theta(v) > 2\pi$ if $val(v) > 6$, $\theta(v) \geq 6 \times \pi/3 = 2\pi$ if $val(v) = 6$. Assuming (1) is not true, then we also have

$$\theta(v) \geq 3 \times \frac{\pi}{3} + 2 \times \frac{\pi}{4} = 2\pi$$

if $val(v) = 5$. Thus there is no vertex with cone angle $\theta(v) < 2\pi$, so by Lemma 11.5 we see that $\theta(v) = 2\pi$ for all $v \in G$, hence there is no vertex of valence more than 6, all faces incident to vertices of valence 6 are 3-gons, and exactly 3 faces incident to a vertex of valence 5 are 3-gons and the other two are 4-gons. Therefore either (2) or (3) holds. □

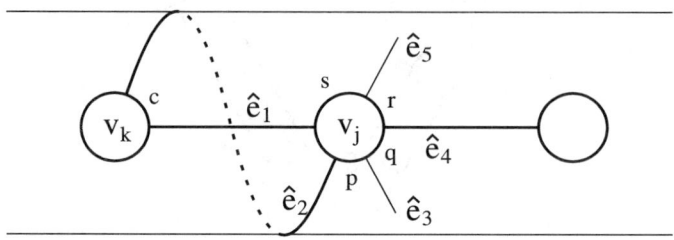

Figure 11.7

LEMMA 11.7. *Suppose $n_a = 4$, $n_b > 2$, and Γ_b is positive. Then no vertex v_j of $\hat{\Gamma}_b$ with $val(v_j) = 5$ has four corners belonging to 3-gons.*

PROOF. Let v_j be a vertex of $\hat{\Gamma}_b$ with $val(v_j) = 5$. By Lemma 11.1(4) we must have $\Delta = 4$. The weights of the edges at v_j are $4, 4, 4, 2, 2$. By Lemma 11.1(2) the three weight 4 edges are not consecutive, hence the order around v_j is $(4, 4, 2, 4, 2)$. Label these edges by $\hat{e}_1, ..., \hat{e}_5$, respectively, where $\hat{e}_3, \hat{e}_5$ have weight 2. By Lemma 11.1(5) the two edges $\hat{e}_1, \hat{e}_2$ form an essential cycle on the torus $\hat{F}_b$, hence the graph looks like that in Figure 11.7.

Let c be the corner at v_k between these two edges $\hat{e}_1, \hat{e}_2$, as shown in Figure 11.7. We claim that c contains no other edge endpoint. Let e and e' be the edges in $\hat{e}_1$ and $\hat{e}_2$ with label 1 at v_j. Let P, Q be the endpoints of e, e' at v_j, and let R, S be the endpoints of e, e' at v_k, respectively. By Lemma 6.5 these edges are parallel on Γ_a, so they connect the same pair of vertices u_1, u_r for some r. On Γ_b this means that P, Q have the same label 1, and R, S have the same label r.

Since e, e' are parallel negative edges on Γ_a, we have $d_{u_1}(P, Q) = d_{u_r}(R, S)$, therefore the four points P, Q, R, S satisfy the assumptions of Lemma 2.16(1), hence by the lemma we have $d_{v_j}(P, Q) = d_{v_k}(R, S)$. Without loss of generality assume that the orientations on $\partial v_j, \partial v_k$ are counterclockwise on Figure 11.7. Then one can see that $d_{v_j}(P, Q) = 4$, hence $d_{v_k}(R, S) = 4$, which implies that there are only 3 edge endpoints from the endpoint of e to that of e' on ∂v_k, so there is no edge endpoint in the corner c in Figure 11.7. This proves the claim.

Label the corners at v_j as shown in Figure 11.7. The above implies that the corners p and s belong to the same face σ, so if v_j is incident to at least four 3-gons then σ must be a 3-gon, hence $\hat{e}_3 = \hat{e}_5$ is a loop. Now the corners q and r belong to the same face σ', which cannot be a 3-gon, hence the result follows. □

LEMMA 11.8. *Suppose $n_a = 4$, $n_b > 2$, and Γ_b is positive. Suppose $\Delta = 4$.*

(1) All faces of $\hat{\Gamma}_b$ are 3-gons or 4-gons, every vertex has valence 5 and has exactly three 3-gons incident, and the weight sequence of the edges incident to the vertex is $(4, 4, 2, 4, 2)$. In particular, n_b is even.

(2) The two weight 2 edges at any vertex form a loop, which is incident to a 3-gon whose other two edges are of weight 4.

(3) Each edge in a weight 2 family of Γ_b has label pair (23) or (14).

11. THE CASE $n_a = 4$, AND Γ_b POSITIVE

PROOF. (1) By Lemma 11.1(3) and Lemma 11.6, $\hat{\Gamma}_b$ is one of the three types stated in Lemma 11.6. Lemma 11.7 shows that $\hat{\Gamma}_b$ cannot be of type (1). If $\hat{\Gamma}_b$ has a vertex of valence 6 then the weights are $4, 4, 2, 2, 2, 2$, hence there are two consecutive edges of weight 2. By Corollary 11.4(3) the three faces incident to these two edges cannot all be 3-gons, hence $\hat{\Gamma}_b$ cannot be of type (2) in Lemma 11.6. It follows that $\hat{\Gamma}_b$ is of type (3) in Lemma 11.6, so the weights of the edges at every vertex of $\hat{\Gamma}_b$ are $4, 4, 4, 2, 2$. Thus the number of weight 4 edge endpoints in $\hat{\Gamma}_b$ is $3n_b$, which must be even, hence n_b is even. By Lemma 11.1(2) the three weight 4 edges cannot all have the same label sequence, hence the weight sequence is $(4, 4, 2, 4, 2)$ at each vertex.

(2) By Lemma 11.1(5) the two adjacent weight 4 edges at v_j connects v_j to a vertex v_k and form an essential loop on $\hat{F}_b$. The other weight 4 edge at v_j connect to some vertex v_r, whose two other weight 4 edges connect to another vertex v_s and form an essential loop. These five weight 4 edges cut off a 6-gon containing the four weight 2 edges at v_j and v_r. The 6-gon cannot contain any vertex in its interior because each vertex is incident to two weight 4 edges forming an essential cycle on $\hat{F}_b$. Therefore the four weight 2 edges at v_j and v_r form two loops.

(3) The loop $\hat{e}$ of $\hat{\Gamma}_b$ at v_j cuts off a 3-gon in the 6-gon above. Let e_1 be the edge of $\hat{e}$ which is on the boundary of a 3-gon face σ of Γ_b. Then the other two edges of σ belong to families of 4 edges and hence must have label pair (12) and (34) respectively, so the labels on $\partial\sigma$ are as shown in Figure 11.3, and e_1 has label pair (23) or (14). Since the other loop edge of $\hat{e}$ is parallel to e_1, it has label pair (14) or (23), respectively. □

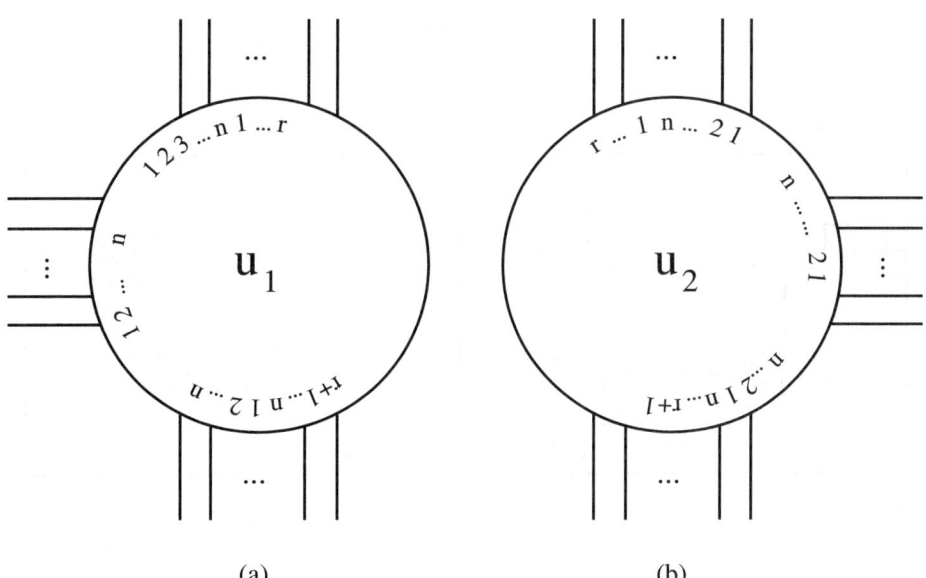

Figure 11.8

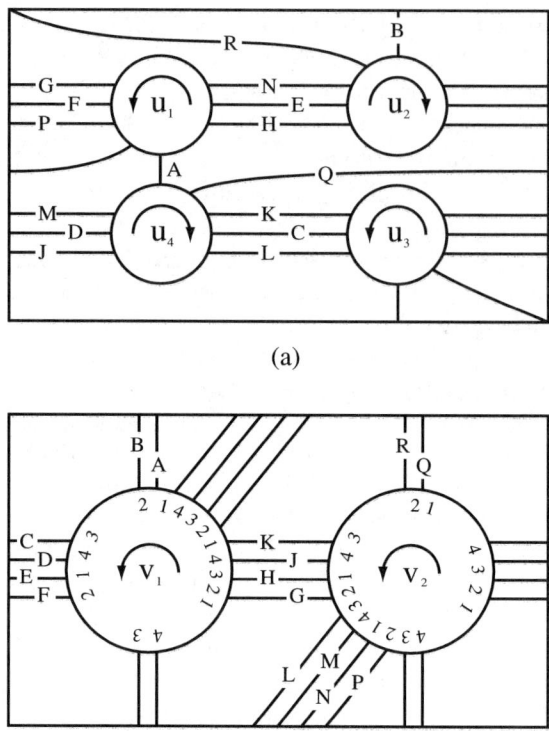

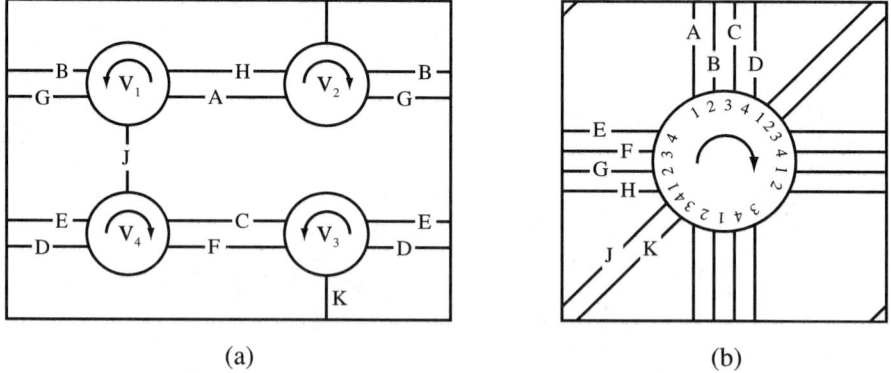

Figure 11.10

PROPOSITION 11.9. *Suppose $n_a = 4$ and Γ_b is positive.*
(1) If $\Delta = 4$ then $n_b = 2$, and the graphs are as shown in Figure 11.9.
(2) If $\Delta = 5$ then $n_b = 1$, and the graphs are as shown in Figure 11.10.

11. THE CASE $n_a = 4$, AND Γ_b POSITIVE

PROOF. (1) Put $n = n_b$. If $n = 1$ then the assumption $\Delta = 4$ implies that the weights of the edges of $\hat{\Gamma}_b$ are either $(4, 4, 0)$ or $(4, 2, 2)$. In either case a family of four edges form an extended Scharlemann cycle, which is impossible.

Now assume $n \geq 2$. Let $C = \hat{e}_1 \cup \hat{e}_2$ be the two edges in $\hat{\Gamma}_a$ connecting u_1 to u_2. Since $\hat{F}_a$ contains both (12)- and (34)-Scharlemann cocycles, any edge of Γ_b with label pair (12) is parallel to an edge of C on Γ_a. By Lemma 11.8(1) each vertex of $\hat{\Gamma}_b$ has valence 5 and hence is incident to 3 weight 4 edges, so $\hat{\Gamma}_b$ has $3n/2$ weight 4 edges. Each weight 4 edge contributes 2 edges to C, one for each $\hat{e}_i$, and by Lemma 11.8(3) no weight 2 edge of $\hat{\Gamma}_b$ contributes to C. Thus each $\hat{e}_i$ represents exactly $3n/2$ edges, and each label j appears exactly three times among the edge endpoints of $\hat{e}_1 \cup \hat{e}_2$ at u_1. Hence if the edge endpoints of $\hat{e}_1$ at u_1 are labeled $1, ..., n, 1, ..., r$, then the labels of those in $\hat{e}_2$ must be $r+1, ..., n, 1, ..., n$, where $r = n/2$. It follows that the n edge endpoints at u_1 that do not belong to C must be on one side of C, so up to relabeling we may assume that the labels at u_1 are as shown in Figure 11.8(a). Let φ be the involution on $\hat{F}_a$ given by Lemma 6.2(4). Then φ maps $\hat{e}_1$ to $\hat{e}_2$ and is label preserving, so the labels at u_2 must be as shown in Figure 11.8(b). Now the transition function of $\hat{e}_1$ maps 1 to $r+1$, which has period 2. Since $n > 2$, this function is not transitive, contradicting Lemma 2.3(1).

We have shown that $n = 2$. The graph $\hat{\Gamma}_b$ is now a subgraph of that in Figure 13.1, with vertices labeled v_1, v_2 instead of u_1, u_2. Let w_i be the weight of $\hat{e}_i$. By Lemma 6.4(2) Γ_a is kleinian, and by Lemma 6.2(2) the w_i are all even. By Lemma 11.1(3) we have $w_5 > 0$. If $w_5 = 4$ then $\hat{e}_5$ containing no extended Scharlemann bigon implies that either $w_1 + w_2 = 6$ or $w_3 + w_4 = 6$, so $w_i = 4$ for some $i \leq 4$, in which case $\hat{e}_i$ contains a (12)-Scharlemann bigon, whose edges, by the above, must be parallel in Γ_a to the (12)-edges of $\hat{e}_5$, which is a contradiction to Lemma 2.3(5). Therefore we must have $w_5 = 2$. For the same reason, the two loops in $\hat{e}_5$ cannot be a Scharlemann bigon, so we must have $w_1 = w_2 = 4$ and $w_3 + w_4 = 4$ up to symmetry. Let $e_1 \cup ... \cup e_4$ and $e'_1 \cup ... \cup e'_4$ be the edges in $\hat{e}_1, \hat{e}_2$ respectively, such that e_i, e'_i have label i at v_1. By Lemma 6.5 e_i, e'_i are parallel on Γ_a, with the same label 1 at u_i. Therefore there is another edge between them, which must belong to $\hat{e}_3 \cup \hat{e}_4$. If $w_3 = w_4 = 2$ then one can check that these edges would have label pairs (14) and (23), which is a contradiction. Therefore we may assume $w_3 = 4$ and $w_4 = 0$. The graph Γ_b is now as shown in Figure 11.9(b).

As shown above, there are 12 edges on Γ_b with label pair (12) or (34), divided into 4 families of 3 edges each on Γ_a. Label the edges of Γ_b as in the figure. Up to symmetry we may assume the edge A is as shown in Figure 11.9(a). Since $\Delta = 4$, we may assume that the jumping number $J = 1$. The 1-edges around v_1 appear in the order A, E, G, P. By Lemma 6.5 G, P are parallel on Γ_a. This determines the position of these edges as well as the orientation of u_1. The 2-edges at v_2 appear in the order E, R, G, P, and E on Γ_a has already been determined above, hence the position of R, G, P must appear around ∂u_2 are shown in the figure. Other edges on Γ_a can be determined similarly.

(2) The proof for $\Delta = 5$ is similar but simpler. In this case by Lemma 11.1(4) each vertex v_j of $\hat{\Gamma}_b$ has valence 6, and the edges at v_j have weights $4, 4, 4, 4, 2, 2$. Thus all white faces are bigons and 3-gons, so by Corollary 11.4(1) $\hat{\Gamma}_a$ is a subgraph of that in Figure 11.5. By Lemma 11.1(2) the two weight 2 edges at any vertex are non-adjacent, thus any edge e in a weight 2 family is on the boundary of a 3-gon whose other two edges have label pairs (12) and (34), so the label pair of e must be

(14) or (23) and hence e is not a vertical edge in Figure 11.5. On the other hand, each weight 4 edge of $\hat{\Gamma}_b$ contributes one edge to each vertical family in Figure 11.5, hence each vertical edge has weight exactly $2n_b$. As above, one can show that the transition function defined by a family of vertical edges is the identity function, which by Lemma 2.3(1) implies that $n_b = 1$, so the graph Γ_b must be as shown in Figure 11.10(b). By the above discussion, Γ_a is as shown in Figure 11.10(a). Label edges as in Figure 11.10(b). By Lemma 6.5 the edges A, H are parallel on Γ_b, hence we may assume the jumping number $J = 1$. One can now easily determine the labels of the edges of Γ_a. □

PROPOSITION 11.10. *Suppose $n_a \leq n_b$. Then $n_a \leq 2$.*

PROOF. By Proposition 5.11 we have $n_a \leq 4$. Assume $n_a = 4 \leq n_b$. Then by Proposition 11.9, Γ_b is non-positive. Therefore, by Proposition 9.7, $n_a = n_b = 4$, and, by Proposition 11.9 again, Γ_a is also non-positive, contradicting Proposition 10.5.

Suppose $n_a = 3$. Then $\hat{F}_a$ is non-separating. If Γ_a is non-positive then some vertex u_1, say, has different sign to the other two vertices u_2, u_3. One of the vertices has valence at most two in $\hat{\Gamma}_a^+$, so it is incident to at most $2n_b$ positive edges, and hence at least $2n_b$ negative edges. By Lemma 2.8 this implies that Γ_b has a Scharlemann cycle, so by Lemma 2.2(4) $\hat{F}_a$ is separating, which is a contradiction. If Γ_a is positive then by Lemma 3.1 we have $n_b \leq 4$. By Lemma 2.23 n_1, n_2 cannot both be odd, hence $n_a = 3$ implies that n_b is even, so we must have $n_b = 4$. Now applying Proposition 11.9 with n_a, n_b reversed, we get $n_a \leq 2$, which is a contradiction. □

12. The case $n_a = 2$, $n_b \geq 3$, and Γ_b positive

The next few sections deal with the case that $n_a = 2$ and $n_b \geq 3$. The main result of this part is Proposition 16.8, which shows that there are only a few possibilities for the graphs Γ_a, Γ_b.

Throughout this section we will assume that $n_a = 2$, $n_b \geq 3$ and Γ_b is positive. We will show that this case is impossible. To simplify notation, denote n_b by n. Note that $\hat{\Gamma}_a$ has at most four edges $\hat{e}_1, ..., \hat{e}_4$, all connecting u_1 to u_2. We will always assume that the first edge of $\hat{e}_1$ has label 1 at u_1. Write $\Gamma_a = (a_1, a_2, a_3, a_4)$, where a_i is the weight of $\hat{e}_i$. Let $s_i = s(\hat{e}_i)$ be the transition number of $\hat{e}_i$ from u_1 to u_2. In the following lemma the subscripts are integers mod 4.

LEMMA 12.1. *(1) $s_{i+1} \equiv s_i - a_i - a_{i+1} \mod n$;*
(2) $s_{i+2} \equiv s_i - a_{i+1} + a_{i+3} \mod n$. In particular, $s_i \equiv s_{i+2}$ if and only if $a_{i+1} \equiv a_{i+3} \mod n$.

PROOF. (1) Orient edges from u_1 to u_2, and denote by $e(h), e(t)$ the head and tail of an edge e. Let e, e' be the first edge of $\hat{e}_i, \hat{e}_{i+1}$, respectively. Let x be the label of $e(t)$, and y the label of $e'(h)$. Then traveling from $e(t)$ to $e'(t)$ on ∂u_1 then to $e'(h)$ through e' gives $y \equiv x + a_i + s_{i+1} \mod n$, while traveling through e to $e(h)$ then along ∂u_2 to $e'(h)$ gives $y \equiv x + s_i - a_{i+1} \mod n$. Hence $s_{i+1} \equiv s_i - a_i - a_{i+1} \mod n$.

(2) Applying (1) twice gives $s_{i+2} \equiv s_i - a_i - a_{i+1} - a_{i+1} - a_{i+2} \equiv s_i - a_{i+1} + a_{i+3} \mod n$. □

LEMMA 12.2. *Let e, e' be edges of Γ_b joining a pair of distinct vertices, such that $e \cup e'$ is null-homotopic in $\hat{F}_b$. If e belongs to a family of at least n parallel edges in Γ_a then e and e' are parallel on Γ_b.*

PROOF. Let D be the disk in $\hat{F}_b$ bounded by $e \cup e'$. The family of n parallel edges of Γ_a containing e gives a set of essential loops on $\hat{F}_b$, corresponding to the orbits of the associated permutation. It follows that D contains no vertices in its interior, and hence e and e' are parallel on Γ_b. □

LEMMA 12.3. *Suppose Γ_a contains bigons $e_1 \cup e_2$ and $e_1' \cup e_2'$, such that e_1, e_1' have label i at u_1 and j at u_2, and e_2, e_2' have label $i+1$ at u_1 and $j+1$ at u_2. Suppose either*

(i) $j \neq i \pm 1$, or

(ii) Γ_a contains a pair of edges e_3, e_3' with the same label pair (r, s), such that $r, s \notin \{i, i+1, j, j+1\}$ and $C_3 = e_3 \cup e_3'$ form an essential loop on $\hat{F}_b$.

Then $C_1 = e_1 \cup e_1'$ and $C_2 = e_2 \cup e_2'$ are inessential on $\hat{F}_b$.

PROOF. (i) Note that $i \neq j$ by the 2-Cycle Lemma 2.14(3). If $j \neq i \pm 1$ then $C_1 \cap C_2 = \emptyset$, so they cannot be essential and yet non-homotopic on $\hat{F}_b$, hence by Lemma 2.20 they must be inessential.

(ii) In this case C_1, C_2 lie in the interior of the annulus obtained from $\hat{F}_b$ by cutting along C_3, which again implies that C_1, C_2 cannot be essential and yet non-homotopic on $\hat{F}_b$. □

An edge is a *border edge* if it is the first or last edge in a family of parallel edges.

LEMMA 12.4. *Suppose $s_k \neq \pm 1$. Then*

(1) $a_k \leq n+1$, and

(2) if $a_k = n+1$ and e' is an edge of $\hat{e}_j$ which has the same label pair as that of a border edge e_1 of $\hat{e}_k$, then e' is a border edge.

PROOF. (1) Assume $a_k \geq n+2$. Label the first $n+2$ edges of $\hat{e}_k$ successively as $e_1, e_2, ..., e_n, e_{n+1}, e_{n+2}$. Let $e_i' = e_{n+i}$. Since $s_k \neq \pm 1$, e_1, e_2, e_1', e_2' satisfy Condition (i) in Lemma 12.3, so $e_1 \cup e_1'$ is an inessential loop on $\hat{F}_b$. By Lemma 12.2 this implies that e_1 and e_1' are parallel on Γ_b, and hence parallel on both graphs, which is a contradiction.

(2) If e' is not a border edge then the bigon $e_1 \cup e_2$ and one of the two bigons containing e' satisfy the assumption of Lemma 12.3(i), hence e' is parallel to e_1 on Γ_b. Similarly, using the bigon between e_n, e_{n+1} and the other bigon containing e' one can show that e' is also parallel to e_{n+1} on Γ_b, hence e_1, e_{n+1} are parallel on both Γ_a and Γ_b, which is again a contradiction. □

LEMMA 12.5. *Let $\hat{e}, \hat{e}'$ be families of at least n parallel edges in Γ_a, and let $i, j, k, l \in \mathbb{Z}_n$ be distinct. Then Γ_a cannot contain both*

(i) ij-edges e_1, e_2, e_3 with $e_1, e_2 \in \hat{e}$ and e_3 non-equidistant with e_1, e_2; and

(ii) kl-edges $e_1', e_2' \in \hat{e}'$.

PROOF. The edges e_1, e_2, e_3 are pairwise non-parallel in Γ_b. Since $e_1, e_2 \in \hat{e}$, no pair of e_1, e_2, e_3 cobounds a disk in $\hat{F}_b$ by Lemma 12.2. Hence e_1', e_2' cobound a disk in $\hat{F}_b$. Since $e_1' \in \hat{e}'$, e_1', e_2' are parallel on Γ_b by Lemma 12.2. This contradicts Lemma 2.2(2). □

Two families $\hat{e}, \hat{e}'$ of Γ_a are *A-conjugate* it there are $e \in \hat{e}$ and $e' \in \hat{e}'$ such that they are anti-parallel on Γ_b when oriented on Γ_a from u_1 to u_2. They are *P-conjugate* if the e, e' above are parallel on Γ_b as oriented edges. They are *conjugate* if they are either A-conjugate or P-conjugate.

LEMMA 12.6. *(1) There exist $\hat{e}_i, \hat{e}_j$ on $\hat{\Gamma}_a$ which are A-conjugate. Moreover, if $a_i < (\Delta - 3)n$ or $a_j < (\Delta - 3)n$ then there is another such pair. (The two pairs may have one family in common.)*

(2) If $\hat{e}_i, \hat{e}_j$ are A-conjugate then $s_i \equiv -s_j \mod n$; moreover, if $a_i \geq n$ or $a_j \geq n$ then $s_i \not\equiv s_j \mod n$.

PROOF. (1) Since there are Δn edges while $\hat{\Gamma}_b$ has at most $3n$ edges, Γ_b has at least $(\Delta - 3)n$ bigons. The two edges of a bigon in Γ_b belong to a pair of A-conjugate families $\hat{e}_i, \hat{e}_j$ on Γ_a. If $a_i < (\Delta - 3)n$ or $a_j < (\Delta - 3)n$ then these families cannot contain all the bigons on Γ_b, hence there must be another A-conjugate pair.

(2) If $\hat{e}_i, \hat{e}_j$ are A-conjugate then by definition there exist $e \in \hat{e}_i$ and $e' \in \hat{e}_j$ which are anti-parallel on Γ_b, hence the label of e' at u_2 is the same as that of e at u_1, and vice versa. Therefore $s_i \equiv -s_j \mod n$.

If we also have $s_i \equiv s_j \mod n$ then $s_i = 0$ or $n/2$, which is a contradiction to the 2-Cycle Lemma 2.14(3). □

LEMMA 12.7. *Let $\hat{e} = e_1 \cup ... \cup e_p$ and $\hat{e}' = e'_1 \cup ... \cup e'_q$ be two families of Γ_a, where $p \leq q$.*

(1) If $\hat{e}$ and $\hat{e}'$ are conjugate then $p \equiv q \mod 2$.

(2) If $\hat{e}$ and $\hat{e}'$ are conjugate and $q \geq p$ then each edge e_r is parallel to the edge e'_{r+c}, where $c = (q-p)/2$; hence the set of edges in $\hat{e}'$ which are parallel to those in $\hat{e}$ lie exactly in the middle of $\hat{e}'$.

(3) If $p + q \equiv 0 \mod 2n$ and $\hat{e}, \hat{e}'$ are adjacent on $\hat{\Gamma}_a$ then they are not A-conjugate.

(4) If $p + q \equiv 0 \mod 2n$, $\hat{e}, \hat{e}'$ are adjacent on $\hat{\Gamma}_a$, and $J \neq \pm 1$ then they are not conjugate.

PROOF. (1) By definition there are edges e_i, e'_j which are parallel in Γ_b.

First consider the case that $\hat{e}, \hat{e}'$ are adjacent. Denote by $e(k)$ the endpoints of e at u_k. We may assume that the first edge e'_1 of $\hat{e}'$ is adjacent to the last edge e_p of $\hat{e}$ on ∂u_1. Then the distance from $e_i(1)$ to $e'_j(1)$ is

$$(5) \qquad d_{u_1}(e_i, e'_j) = j + p - i$$

On ∂u_2 $e'_q(2)$ is adjacent to $e_1(2)$, so we have

$$(6) \qquad d_{u_2}(e'_j, e_i) = i + q - j$$

Since e_i, e'_j are parallel positive edges on Γ_b, they are equidistant, hence by Lemma 2.17 we have $d_{u_1}(e_i, e'_j) = d_{u_2}(e'_j, e_i)$, which gives

$$(7) \qquad 2(j - i) = q - p$$

and

$$(8) \qquad 2d = p + q$$

where $d = d_{u_1}(e_i, e'_j) = d_{u_2}(e'_j, e_i)$. Equation (7) gives $q - p \equiv 0 \mod 2$.

Now suppose $\hat{e}, \hat{e}'$ are not adjacent. Let $\hat{e}''$ be the family whose endpoints on ∂u_1 are between $e_p(1)$ and $e'_1(1)$. Then on ∂u_2 the endpoints of $\hat{e}''$ are also exactly the ones between $e'_q(2)$ and $e_1(2)$. Thus if $\hat{e}''$ has k edges then the equations (5) and (6) above become $d = j + k + p - i$ and $d = i + k + q - j$. Therefore again we have $2(j - i) = q - p$, and the result follows.

(2) From equation (7) we have $j = i + (q-p)/2$. If $i > 1$ then the above and the condition $q \geq p$ imply that $j > 1$. Since e_i is parallel to e'_j on Γ_b, by Lemma 2.20 applied to the bigons $e_{i-1} \cup e_i$ and $e'_{j-1} \cup e'_j$, the loop $e_{i-1} \cup e'_{j-1}$ is null-homotopic on $\hat{F}_b$, hence by Lemma 12.2 e_{i-1} is parallel to e'_{j-1} on Γ_b. Similarly, if $i < p$ then the edge e_{i+1} is parallel to e'_{j+1} on Γ_b. By induction it follows that every edge e_k in $\hat{e}$ is parallel to the edge $e'_{k+(q-p)/2}$.

(3) Assume without loss of generality that $\hat{e} = \hat{e}_1$ and $\hat{e}' = \hat{e}_2$. If they are A-conjugate then the label of e'_j at u_1 is the same as the label of e_i at u_2, so $d = d_{u_1}(e_i, e'_j) \equiv s_1 \mod n$. Hence equation (8) and the assumption $p + q \equiv 0 \mod 2n$ gives $s_1 \equiv d \equiv 0 \mod n$. Similarly $s_2 \equiv 0 \mod n$. Since $n \geq 3$ and one of p, q is at least n, this is a contradiction to Lemma 2.14(2).

(4) By (3) $\hat{e}, \hat{e}'$ are not A-conjugate. Assume they are P-conjugate.

If $p + q = 4n$ then by Lemma 2.22(3) we have $p = q = 2n$, and by (2) each e_i of $\hat{e}_1$ is parallel to the corresponding edge e'_i of $\hat{e}_2$ on Γ_b for $i = 1, ..., 2n$. Since $\hat{e}, \hat{e}'$ are not A-conjugate, e_i, e'_i are parallel as oriented edges, with orientation from u_1 to u_2. Hence there is another edge e''_i between them, which cannot belong to $\hat{e} \cup \hat{e}'$ as otherwise there would be two edges parallel on both graphs, contradicting Lemma 2.2(2). This gives at least $6n$ edges on Γ_a, which is a contradiction.

Now assume $p + q = 2n$. Let e, e' be the edges of $\hat{e}, \hat{e}'$ which are parallel on Γ_b as oriented edges, so they have the same label k at u_1 for some k. The condition $p + q = 2n$ implies that e, e' are adjacent among edges labeled k at u_1. Since $J \neq \pm 1$, they are non-adjacent on Γ_b among edges labeled 1 at v_k, hence they belong to a family of at least 5 parallel edges, which is a contradiction to Lemma 2.2(2) because $\hat{\Gamma}_a$ has at most 4 edges. $\square$

LEMMA 12.8. *If the jumping number $J = \pm 1$ (in particular if $\Delta = 4$), then Γ_a has at most $n + 1$ parallel edges.*

PROOF. Suppose for contradiction that $\hat{e}_1$, say, contains edges $e_1, ..., e_{n+2}$. By Lemma 12.4(1) we may assume that $s_1 = 1$, so the label sequences of these edges are $(1, 2, ..., n, 1, 2)$ at u_1, and $(2, 3, ..., n, 1, 2, 3)$ at u_2. By Lemma 2.22(1) we may assume that the subgraph of Γ_b consisting of these edges is as shown in Figure 2.3. Up to symmetry we may assume that the orientation of ∂v_i is counterclockwise on Figure 2.3.

Orient edges from u_1 to u_2. Denote by h_i, t_i the head and tail of e_i, respectively. For $i > 1$, h_{i-1} and t_i both have label i on Γ_a, so they are on ∂v_i. Define $d_i = d_{v_i}(t_i, h_{i-1})$, where $i = 2, ..., n + 2$. Note that $d_i = 1$ implies that the corner from t_i to h_{i-1} on ∂v_i contains no edge endpoint.

CLAIM 1. $d_i = d_j$ for $2 \leq i, j \leq n + 2$.

PROOF. Isotoping on T_0 along the positive direction of ∂u_i moves h_1 to h_2 and t_2 to t_3, so the distance on ∂v_2 from h_1 to t_2 should be the same as that on ∂v_3 from h_2 to t_3, i.e., $d_2 = d_3$. (Alternatively one may apply Lemma 2.16 to obtain the result.) Similarly we have $d_i = d_{i+1}$ for $2 \leq i \leq n + 1$. $\square$

CLAIM 2. $d_i = 1$ for $2 \leq i \leq n + 2$.

PROOF. By assumption we have $J = \pm 1$, so either $d_{v_2}(h_1, h_{n+1}) = 2$ (when $J = 1$), or $d_{v_1}(t_{n+1}, t_1) = 2$ (when $J = -1$). In the first case, from Figure 2.3 we see that the tail of e_{n+2} is the only edge endpoint at the corner from h_1 to h_{n+1}, hence $d_{n+2} = 1$. Similarly in the second case the head of e_n is the only edge endpoint on ∂v_1 from t_{n+1} to t_1, hence $d_{n+1} = d_{v_1}(t_{n+1}, h_n) = 1$. In either case by Claim 1 we have $d_i = 1$ for all i between 2 and $n+2$. □

Let D be the disk face indicated in Figure 2.3. Claim 2 shows that all corners of D except c_1, c_2, c_3 shown in the figure contain no edge endpoints.

When $J = 1$, we have $d_{v_1}(t_1, t_{n+1}) = 2$, so there is one edge endpoint in c_1. Similarly there is one edge endpoint in c_3. Since $d_{v_2}(t_{n+2}, t_2) = 2\Delta - 2 \geq 6$, there are at least 4 edge endpoints in c_2. Thus there would be some trivial loops based at v_2, contradicting the assumption that Γ_b has no trivial loops.

When $J = -1$, $d_{v_1}(t_1, t_{n+1}) = d_{v_3}(h_2, h_{n+2}) = 2\Delta - 2 \geq 6$, and $d_{v_2}(t_{n+2}, t_2) = 2$, so there are at least 5 edge endpoints in each of c_1 and c_3, and no edge endpoints in c_2. It follows that D contains at least 5 interior edges, all parallel to each other, two of which would then be parallel on both graphs, contradicting Lemma 2.2(2). □

LEMMA 12.9. *Γ_a has at most $n + 2$ parallel edges.*

PROOF. Assume to the contrary that $\hat{e}_1 \supset e_1 \cup ... \cup e_{n+3}$. By Lemma 12.4(1) we may assume without loss of generality that $s_1 = 1$. By Lemma 2.22(1) the first $n+2$ edges appear in Γ_b as shown in Figure 2.3

First assume $n \geq 4$. Orient edges of Γ_a from u_1 to u_2, and denote by $e(h), e(t)$ the head and tail of an edge e, respectively. From Figure 2.3 we see that the triple $(e_2(h), e_{n+2}(h), e_3(t))$ is positive, hence the triple $(e_3(h), e_{n+3}(h), e_4(t))$ is also positive by Lemma 2.21(2), so the head of e_{n+3} lies on the inner circle in Figure 2.3. Note that e_{n+2} shields this edge endpoint from the outside circle of the annulus in Figure 2.3, hence the tail of e_{n+3} also lies in the inner circle in the figure, therefore e_{n+3} is parallel to e_3 on Γ_b, which is a contradiction as they cannot be parallel on both graphs.

Now consider the case $n = 3$. By Lemma 2.22(3) we may assume $a_i \leq 6$, and $a_1 = n+3 = 6$. By Lemma 12.8 we may assume that $\Delta = 5$ and the jumping number $J \neq \pm 1$. Also, $a_j \neq 5$, otherwise by Lemma 12.7(1) the 11 edges in $\hat{e}_1 \cup \hat{e}_j$ would be mutually non-parallel on Γ_b, contradicting the fact that $\hat{\Gamma}_b$ has at most $3n = 9$ edges (Lemma 2.5). One can now check that the following are the only possible values of (a_1, a_2, a_3, a_4) up to symmetry, where $*$ indicates any possible value. Let $s = s_1$. Then the other s_i can be calculated using Lemma 12.1. The second quadruple in the following list indicates the values of (s_1, s_2, s_3, s_4).

(1)	$(6, 6, *, *)$	$(s, s, *, *)$
(2)	$(6, 1, 6, 2)$	$(s, s-1, s+1, s-1)$
(3)	$(6, 1, 4, 4)$	$(s, s-1, s, s+1)$
(4)	$(6, 4, 3, 2)$	$(s, s-1, s+1, s-1)$
(5)	$(6, 4, 1, 4)$	$(s, s-1, s, s+1)$
(6)	$(6, 3, 2, 4)$	$(s, s, s+1, s+1)$
(7)	$(6, 3, 4, 2)$	$(s, s, s-1, s-1)$
(8)	$(6, 3, 3, 3)$	(s, s, s, s)

In case (1) by Lemma 12.7(4) the 12 edges in $\hat{e}_1 \cup \hat{e}_2$ are mutually non-parallel on Γ_b, which is impossible because $\hat{\Gamma}_b$ contains at most $3n = 9$ edges. Case (8) is impossible by Lemma 12.6. Also, Lemma 2.14(3) implies that $s_i \not\equiv 0 \mod 3$ if $a_i \geq 2$, which can be applied to exclude cases (2), (4) and (5).

In case (6) and (7), by Lemma 12.7(1) $\hat{e}_2$ is not conjugate to $\hat{e}_1, \hat{e}_3$ or $\hat{e}_4$, so $\hat{e}_1 \cup \hat{e}_2$ represents all 9 edges in $\hat{\Gamma}_b$, hence each of $\hat{e}_3$ and $\hat{e}_4$ must be conjugate to $\hat{e}_1$. By Lemma 12.7(2) the two middle edges of $\hat{e}_1$ are parallel to the middle edges in each of $\hat{e}_3$ and $\hat{e}_4$, so $\hat{e}_3, \hat{e}_4$ are conjugate. Since $a_3 + a_4 = 6 = 2n$ and $J \neq \pm 1$, this is a contradiction to Lemma 12.7(4).

In case (3), by Lemma 2.14(3) we have $s = 1$. Since $a_1 + a_3 = a_1 + a_4 = 10$ while $\hat{\Gamma}_b$ has at most 9 edges, each of $\hat{e}_3, \hat{e}_4$ must have an edge parallel to some edge of $\hat{e}_1$ on Γ_b. By Lemma 12.7(2) this implies that each edge of $\hat{e}_3 \cup \hat{e}_4$ is parallel to one of the 4 middle edges in $\hat{e}_1$. Note that the edge e' in $\hat{e}_2$ is a loop based at v_1 in Γ_b, which cannot be parallel to any other edge on Γ_b. Therefore Γ_b has exactly 7 families. Moreover, if we let e_1, e_6 be the first and last edges in $\hat{e}_1$ then each of e_1, e_6, e' forms a single family.

Now consider the graph in Figure 2.4. Clearly there is only one possible position for e', which has exactly one endpoint on the corner from the tail of e_1 to the head of e_6. By the above there are no other edge endpoints in this corner, which is a contradiction because the label of the tail of e_1 is 1 while the label of the head of e_6 is 2, so the number of edge endpoints between them must be even. □

LEMMA 12.10. *Suppose $n \geq 4$. Then $\Delta = 4$ and Γ_a has at most $n + 1$ parallel edges.*

PROOF. We need only show that $\Delta = 4$. The second statement will then follow from Lemma 12.8.

Suppose to the contrary that $\Delta = 5$. First assume that $a_i < n + 2$ for all i. Then $\Delta n = 5n \leq 4(n+1)$, so $n = 4$, and $\Gamma_a = (5, 5, 5, 5)$. By Lemma 2.3(1) s_1 is coprime with $n = 4$, so we may assume without loss of generality that $s_1 = 1$. Thus the label sequences of $\hat{e}_1$ are $(1, 2, 3, 4, 1)$ at u_1 and $(2, 3, 4, 1, 2)$ at u_2. One can check that the label sequences of $\hat{e}_3$ are $(3, 4, 1, 2, 3)$ at u_1, and $(4, 1, 2, 3, 4)$ at u_2. This contradicts Lemma 12.5 with e_1, e_2 the two 12-edges in $\hat{e}_1$, e_3 the 12-edge in $\hat{e}_3$, and e'_1, e'_2 the two 34-edges in $\hat{e}_3$.

We may now assume without loss of generality that $a_1 > n + 1$. By Lemma 12.9 we must have $a_1 = n + 2$. By Lemma 12.4(1) we have $s_1 = \pm 1$.

CLAIM 1. $a_2, a_4 \leq n + 1$.

PROOF. Suppose $a_2 = n + 2$. Then $s_2 = \pm 1$ by Lemma 12.4(1). Also by Lemma 12.1 we have
$$s_1 - s_2 \equiv a_1 + a_2 \equiv 4 \mod n$$
Hence either $n = 4$ and $s_1 = s_2$ ($= 1$ say), or $n = 6$, $s_1 = -1$ and $s_2 = 1$. In either case one can check that there is a pair of parallel 12-edges e_1, e_2 in $\hat{e}_1$, a 12-edge e_3 in $\hat{e}_2$ which is not equidistant to e_1, e_2, and a pair of parallel 34-edges e'_1, e'_2 in $\hat{e}_2$. This is a contradiction to Lemma 12.5.

Hence $a_2 \leq n + 1$. A symmetric argument shows that $a_4 \leq n + 1$. □

We now have $5n \leq 2(n+2) + 2(n+1)$, giving $n \leq 6$. Also if $n = 6$ then $\Gamma_a = (8, 7, 8, 7)$.

CLAIM 2. $n = 5$.

PROOF. Otherwise we have $n = 4$ or 6. If $a_2 = n + 1$ then Lemma 12.1 gives $s_2 = s_1 - a_1 - a_2 = \pm 1 - (n+1) - (n+2) \equiv 0 \mod 2$, which is a contradiction to the fact that the transition function of a family of more than n edges must be transitive (Lemma 2.3(1)). Therefore we have $a_2 \leq n$. Similarly for a_4. This rules out the case $n = 6$.

When $n = 4$ we must have $\Gamma_a = (6, 4, 6, 4)$. Assume without loss of generality that $s_1 = 1$. We now apply Lemma 12.5 with e_1, e_2 the 12-edges in $\hat{e}_1$, e_3 the 12-edge in $\hat{e}_3$, and e'_1, e'_2 the 34-edges in $\hat{e}_3$. □

CLAIM 3. If $n = 5$ then $a_3 \neq 7$.

PROOF. Otherwise by Claim 1 we have $(a_2, a_4) = (6, 5)$ or $(5, 6)$, so $a_4 - a_2 = \pm 1$. We may assume that $s_1 = 1$. By Lemma 12.1 we have

$$s_3 \equiv s_1 - a_2 + a_4 = 1 \mp 1 = \ 0 \text{ or } 2 \mod 5$$

which contradicts the fact that $s_3 = \pm 1 \mod n$ (Lemma 12.4(1)). □

The only possibility left is that $n = 5$ and $\Gamma_a = (7, 6, 6, 6)$. We may assume $s_1 = 1$. Then this can be ruled out by applying Lemma 12.5 with e_1, e_2 the 12-edges in $\hat{e}_1$, e_3 the 12-edge in $\hat{e}_3$, and e'_1, e'_2 the 45-edges in $\hat{e}_3$. □

LEMMA 12.11. *(a)* Γ_a *has at most* $n + 1$ *parallel edges.*
(b) $\Delta = 4$.

PROOF. (a) This follows from Lemmas 12.8 and 12.10 if either $J = \pm 1$, or $\Delta = 4$, or $n \geq 4$. Hence we may assume that $\Delta = 5$, $J \neq \pm 1$, and $n = 3$. By Lemma 12.9 we have $a_i \leq n + 2 = 5$. Thus the possible values of (a_1, a_2, a_3, a_4) are given below. The second quadruple gives (s_1, s_2, s_3, s_4), calculated as functions of $s = s_1$, using Lemma 12.1.

(1)	$(5, 5, 5, 0)$	$(s, s-1, s+1, -)$
(2)	$(5, 5, 4, 1)$	$(s, s-1, s-1, s)$
(3)	$(5, 3, 2, 5)$	$(s, s+1, s-1, s+1)$
(4)	$(5, 4, 5, 1)$	(s, s, s, s)
(5)	$(5, 4, 4, 2)$	$(s, s, s+1, s+1)$
(6)	$(5, 4, 3, 3)$	$(s, s, s-1, s-1)$
(7)	$(5, 4, 2, 4)$	(s, s, s, s)
(8)	$(5, 3, 5, 2)$	$(s, s+1, s-1, s+1)$
(9)	$(5, 3, 4, 3)$	$(s, s+1, s, s-1)$

Cases (1), (3), (8) and (9) are impossible because there is an i such that $a_i \geq 2$ and $s_i = 0$, contradicting Lemma 2.14(3). Cases (4) and (7) contradict Lemma 12.6.

In case (2), by Lemma 12.7(1) the edges in $\hat{e}_3$ are not parallel to those in $\hat{e}_1 \cup \hat{e}_4$ on Γ_b, and by Lemma 12.7(4) the edge in $\hat{e}_4$ is not parallel to those in $\hat{e}_1$. Thus the 10 edges in $\hat{e}_1 \cup \hat{e}_3 \cup \hat{e}_4$ are mutually non-parallel on Γ_b, contradicting the fact that $\hat{\Gamma}_b$ has at most $3n$ edges (Lemma 2.5). Similarly, in case (5) the edges in $\hat{e}_1 \cup \hat{e}_3 \cup \hat{e}_4$ are mutually non-parallel on Γ_b, and in case (6) the edges in $\hat{e}_2 \cup \hat{e}_3 \cup \hat{e}_4$ are mutually non-parallel on Γ_b, which lead to the same contradiction.

(b) Assume $\Delta = 5$. By Lemma 12.10 we have $n = 3$, and by (a) we have $a_i \leq 4$, hence the weights of $\hat{e}_i$ must be $(4, 4, 4, 3)$ up to symmetry, and the transition numbers are $(s, s+1, s-1, s+1)$. This is a contradiction to Lemma 2.14(2) because one of the families has $s_i = 0$ and hence is a set of co-loops. □

LEMMA 12.12. *Suppose* $\Delta = 4$. *Let* e, e' *be edges of* Γ_a *with label* i *at vertex* u_1 *and* j *at* u_2, $i \neq j$, *where the i-labels of e, e' at u_1 are not adjacent among all i-labels at u_1. Suppose also that e belongs to a family of at least n parallel edges of Γ_a. Then $e \cup e'$ forms an essential loop on the torus $\hat{F}_b$.*

PROOF. Note that in this case the jumping number $J = \pm 1$, so the assumption that the i-labels of e, e' at u_1 are not adjacent implies that the 1-labels of e_1, e'_1 at the vertex v_i in Γ_b are not adjacent among all 1-labels. By assumption e belongs to a family of at least n parallel edges of Γ_a, so if $e \cup e'$ is inessential on $\hat{F}_b$ then by Lemma 12.2 e_1 and e'_1 are parallel on Γ_b, which gives rise to at least 5 parallel edges in Γ_b, contradicting Lemma 2.2(2) because $\hat{\Gamma}_a$ has at most 4 edges. □

Up to symmetry we may assume that $a_1 \geq a_3$, $a_2 \geq a_4$, and $a_1 + a_3 \geq a_2 + a_4$. Since $a_i \leq n+1$, the possibilities for Γ_a are listed below. The second quadruple indicates the values of s_i, calculated in terms of $s = s_1$ using Lemma 12.1.

(1) $(n+1, n+1, n+1, n-3)$ $(s, s-2, s-4, s-2)$
(2) $(n+1, n+1, n, n-2)$ $(s, s-2, s-3, s-1)$
(3) $(n+1, n+1, n-1, n-1)$ $(s, s-2, s-2, s)$
(4) $(n+1, n, n+1, n-2)$ $(s, s-1, s-2, s-1)$
(5) $(n+1, n, n, n-1)$ $(s, s-1, s-1, s)$
(6) $(n+1, n, n-1, n)$ $(s, s-1, s, s+1)$
(7) $(n+1, n-1, n+1, n-1)$ (s, s, s, s)
(8) (n, n, n, n) (s, s, s, s)

LEMMA 12.13. *Cases (4), (5), (6), (7), (8) are impossible.*

PROOF. In case (4) $\hat{e}_1, \hat{e}_3$ are not A-conjugate to $\hat{e}_2, \hat{e}_4$ by Lemma 12.7(1). Since $a_4 < n$, by Lemma 12.6(1) $\hat{e}_2, \hat{e}_4$ cannot be the only A-conjugate pair, hence $\hat{e}_1$ must be A-conjugate to $\hat{e}_3$. Since they have the same number of edges, by Lemma 12.7(2) the first edge e of $\hat{e}_1$ is parallel to the first edge e' of $\hat{e}_3$. Since e, e' have labels 1, 2 at u_1, respectively, the label of e at u_2 is 2, hence $s = 1$. Now $\hat{e}_2$ is a family of at least 3 co-loops, contradicting Lemma 2.14(2).

In case (5), by Lemma 12.7(1) $\hat{e}_1$ can only be conjugate to $\hat{e}_4$ and $\hat{e}_2$ to $\hat{e}_3$, but since $a_4 < n$, by Lemma 12.6(1) $\hat{e}_2$ must be A-conjugate to $\hat{e}_3$. Since $a_2 + a_3 = 2n$, this is a contradiction to Lemma 12.7(3).

For the same reason, in case (6) $\hat{e}_2$ must be A-conjugate to $\hat{e}_4$. By Lemma 12.7(2) the first edge e_1 of $\hat{e}_2$ must be parallel to the first edge e'_1 of $\hat{e}_4$. Examining the labeling we see that they have labels 2 and 1 at u_1, respectively, so the label of e_1 at u_2 is 1, hence $s_2 = -1$. It follows that $s_1 = s = 0$, which is a contradiction to Lemma 2.3(1).

Cases (7) and (8) are impossible by Lemma 12.6. □

LEMMA 12.14. *Case (1) is impossible.*

PROOF. Since $a_4 < n$, by Lemma 12.6(1) two of the first three families are A-conjugate. Up to symmetry we may assume that $\hat{e}_3$ is A-conjugate to $\hat{e}_2$ or $\hat{e}_1$. By Lemma 12.7(2) the first edges of the above conjugate pair must have the same label pair. Examining the labels of these edges on u_1 we see that $s = 3$ if $\hat{e}_3$ is A-conjugate to $\hat{e}_2$, and $s = 2$ if $\hat{e}_3$ is A-conjugate to $\hat{e}_1$. The second case cannot happen because then $\hat{e}_2$ would be a set of at least 3 co-loops, contradicting Lemma 2.14(2).

The graph Γ_a is now shown in Figure 12.1. If $n \geq 5$ then there are bigons $e_1 \cup e_2$ in $\hat{e}_2$ and e'_1, e'_2 in $\hat{e}_4$ with labels 4, 5 at u_1 and 5, 6 at u_2 (6 = 1 when $n = 5$). Note also that there is a pair of parallel 23-edges $e_3 \cup e'_3$ in $\hat{e}_2$. By Lemma 12.3(ii), these conditions imply that $e_1 \cup e'_1$ is inessential on $\hat{F}_b$, which contradicts Lemma 12.12.

When $n = 4$, there is a pair of 14-edges e_1, e_2 in $\hat{e}_1$, a 14-edge e_3 in $\hat{e}_4$, and a pair of 23-edges in $\hat{e}_2$. Note that e_3 is not equidistant to e_1, e_2. This leads to a contradiction to Lemma 12.5.

When $n = 3$, $s_i = 0$ for some $i = 1, 2, 3$, so one of the first three families contains 4 co-loop edges, which is a contradiction to the 3-Cycle Lemma. □

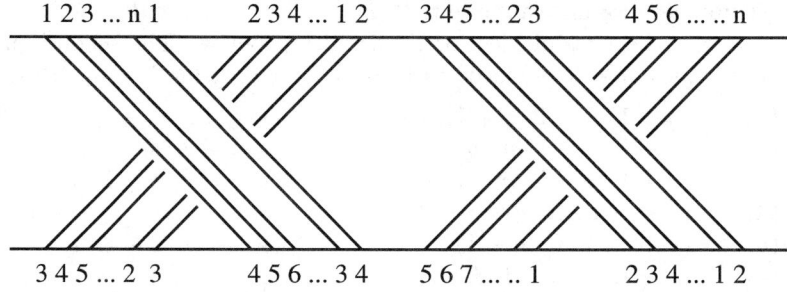

Figure 12.1

LEMMA 12.15. *Case (2) is impossible.*

PROOF. In this case $a_1 \equiv a_2 \not\equiv a_3 \equiv a_4$ mod 2, so by Lemma 12.7(1) no edge in $\hat{e}_1 \cup \hat{e}_2$ is parallel to an edge in $\hat{e}_3 \cup \hat{e}_4$. Since Γ_b contains at least n bigons while $\hat{e}_3 \cup \hat{e}_4$ contributes at most $a_4 = n - 2$ bigons on Γ_b, it follows that some edge in $\hat{e}_2$ is parallel to an edge in $\hat{e}_1$ on Γ_b. Since $a_1 = a_2$, by Lemma 12.7(2) this implies that the first edge e_1 of $\hat{e}_1$ is parallel to the first edge e'_1 of $\hat{e}_2$ on Γ_b. In particular, they must have the same label pair. Since e_1 has label 1 at u_1 and e'_1 has label 2 at u_1, we see that e_1 has label 2 at u_2, hence $s = 1$. Since $s_4 = s - 1 = 0$, this is a contradiction to Lemma 2.14(3) unless $a_4 = n - 2 < 2$, i.e. $n \leq 3$.

Now suppose $n = 3$. Let e_1, e_2 be the two 12-edges in $\hat{e}_1$. Note that there is a 12-edge e_3 in $\hat{e}_3$, which by the above is not parallel to any edge in $\hat{e}_1$, hence e_1, e_2, e_3 cut $\hat{F}_b$ into a disk. Now $\hat{e}_4$ is a loop based at v_3 in Γ_b, so it must be a trivial loop. This is a contradiction because Γ_b contains no trivial loop. □

LEMMA 12.16. *Case (3) is impossible.*

PROOF. We claim that $s = 1$. By Lemma 12.6(1) one of $\hat{e}_1, \hat{e}_2$ is A-conjugate to some other $\hat{e}_j$. Because of symmetry we may assume that $\hat{e}_1$ is conjugate to some $\hat{e}_j$. If $j = 2$ then by Lemma 12.7(2) the first edge e_1 of $\hat{e}_1$ is parallel on Γ_b to the first edge e'_1 of $\hat{e}_2$, which has label 2 at u_1, hence e_1 has label 2 at u_2, so $s = s_1 = 1$. Similarly, if $j = 3$ then by Lemma 12.7(2), $a_1 = n + 1$ and $a_3 = n - 1$ implies that the second edge e^2 of $\hat{e}_1$ is parallel on Γ_b to the first edge of $\hat{e}_3$, which has label 3 at u_1, hence e^2 has label pair (23), which again implies that $s = 1$. The case $j = 4$ is impossible by Lemma 12.6(2). The graph Γ_a is now shown in Figure 12.2.

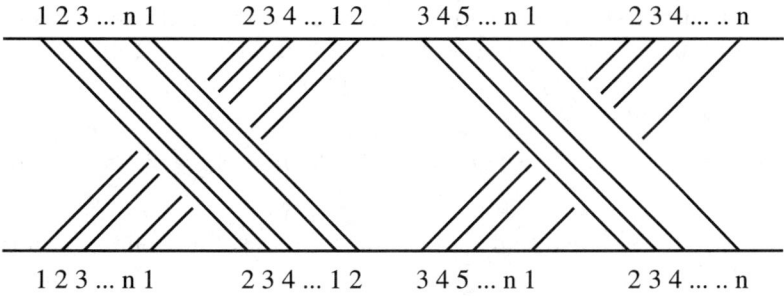

Figure 12.2

There are four edges $e'_1, ..., e'_4$ with label pair (23), where $e'_i \in \hat{e}_i$. One can check on Figure 12.2 that they are all equidistant to each other. We claim that they are all parallel in Γ_b.

The first n edges of $\hat{e}_1$ form a loop C on $\hat{F}_b$. Let a_1, a_2, a_3 be the first three edges of $\hat{e}_1$, oriented from u_1 to u_2, and let $a_i(t), a_i(h)$ be the tail and head of a_i, respectively. Then as in the proof of Lemma 2.22(1), one can show that $d_{v_2}(a_1(h), a_2(t)) = d_{v_3}(a_2(h), a_3(t))$. In other words, the corners at v_2, v_3 on one side of the above loop contain the same number of edge endpoints. Since e'_2 is equidistant to $e'_1 = a_2$, we have $d_{v_2}(a_2(t), e'_2(h)) = d_{v_3}(e'_2(t), a_2(h))$, hence the two endpoints of e'_2 lie on the same side of the loop C. It follows that e'_2 is parallel to e'_1. Similarly, e'_3, e'_4 are also parallel to e'_1. This proves the claim above.

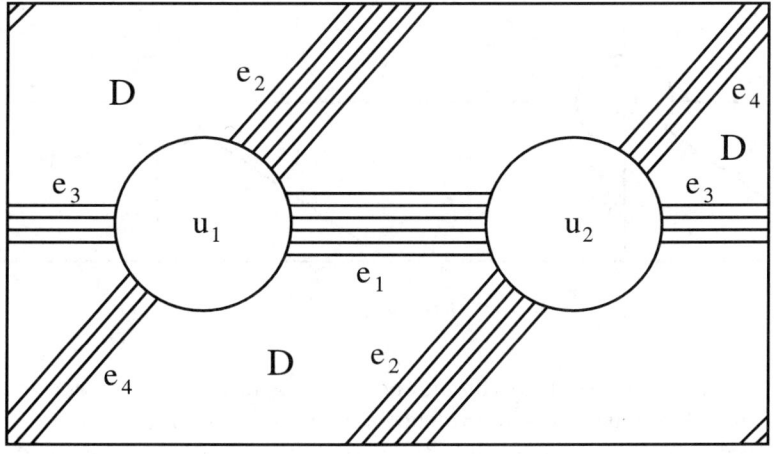

Figure 12.3

Among the four parallel edges $e'_1, ..., e'_4$, at least one of e'_3, e'_4 is adjacent to e'_1 or e'_2 on Γ_b. Because of symmetry we may assume without loss of generality that e'_3 is adjacent to e'_1 or e'_2. Relabel it as e_3.

Note that e_3 is a border edge. There is a face D of Γ_a with $\partial D = e_1 \cup e_2 \cup e_3 \cup e_4$, see Figure 12.3. Let α be the arc in D connecting the middle points of e_2, e_4. Since $e_3 = e'_3$ is parallel and adjacent to e'_1 or e'_2 and e'_1, e'_2 are non-border edges in Γ_a, the face D has a bigon as a coupling face. (See Section 2 for definition.) It follows from Lemma 2.15 that the surface F_a can be isotoped rel ∂ so that the new intersection graph Γ'_a is obtained from Γ_a by deleting e_2, e_4 and replacing them with two edges parallel to e_1, e_3 respectively. The first family of Γ'_a has $n+2$ edges, which is a contradiction to Lemma 12.8. □

PROPOSITION 12.17. *The case that $n_a = 2$, $n = n_b \geq 3$ and Γ_b positive, is impossible.*

PROOF. We have shown that Γ_a has 8 possibilities. These have been ruled out in Lemmas 12.13 – 12.16. □

13. The case $n_a = 2$, $n_b > 4$, Γ_1, Γ_2 non-positive, and $\max(w_1 + w_2, w_3 + w_4) = 2n_b - 2$

It has been shown in Section 12 that if $n_a \leq 2$ and $n_b \geq 4$ then Γ_b cannot be positive. In Sections 13–16 we will discuss the case that Γ_b is non-positive. The results will be given in Propositions 14.7 and 16.8.

As before, we will use n to denote n_b.

LEMMA 13.1. *Suppose $n_a = 2$, $n \geq 4$, and Γ_1, Γ_2 are non-positive.*
(1) The reduced graph $\hat{\Gamma}_a$ is a subgraph of the graph shown in Figure 13.1.
(2) Let w_i be the weight of $\hat{e}_i$. Then up to relabeling we may assume $w_3 + w_4 \leq w_1 + w_2$, and $w_1 + w_2 = 2n - 2$ or $2n$.

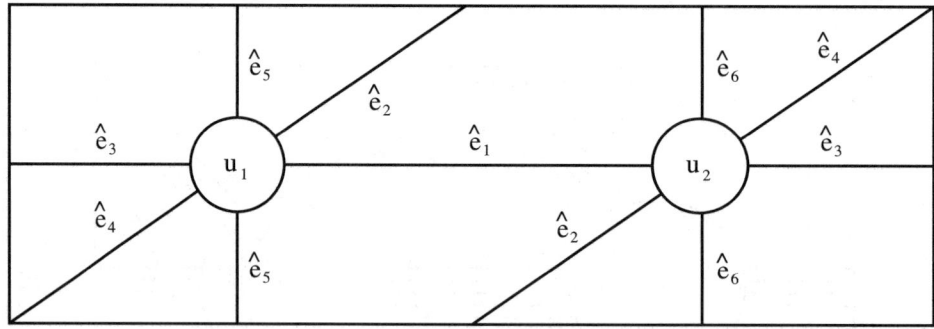

Figure 13.1

PROOF. (1) First note that the number of loops in Γ_a at u_1 is the same as that at u_2, because they have the same valence and the same number of non-loop edges. If $\hat{\Gamma}_a$ has two loops based at u_1 then they cut the torus into a disk, so there is no loop at u_2, which would be a contradiction to the above. Therefore $\hat{\Gamma}_a$ has at most one loop edge at each vertex. If there is no loop at u_i then $\hat{\Gamma}_a$ has at most four edges connecting u_1 to u_2. If there is one loop of $\hat{\Gamma}_a$ at each u_i then these cut the torus into two annuli, each containing at most two edges of $\hat{\Gamma}_a$. In either case $\hat{\Gamma}_a$ is a subgraph of that in Figure 13.1.

13. THE CASE $n_a = 2$, $n_b > 4$, Γ_1, Γ_2 NON-POSITIVE, I

(2) Up to relabeling we may assume that $w_1 + w_2 \geq w_3 + w_4$. Since Γ_b is non-positive, by Lemma 2.3(1) we have $w_i \leq n$, hence $w_1 + w_2 \leq 2n$.

Assume $w_1 + w_2 = 2n - k$ and $k > 0$. Then $w_5 \geq (4n - (w_1 + ... + w_4))/2 \geq k$. Note that the set of k edges $e_1 \cup ... \cup e_k$ of $\hat{e}_5$ adjacent to $\hat{e}_1 \cup \hat{e}_2$ has the same set of labels at each of its two ends. Hence by Lemma 2.4 $\hat{e}_5$ contains a Scharlemann bigon, so by Lemma 2.2(4) and the parity rule k must be even. If $k \geq 4$ then $e_1 \cup ... \cup e_k$ contains an extended Scharlemann bigon, which is a contradiction to Lemma 2.2(6). Hence $k = 2$. □

If Γ_b has a Scharlemann cycle then by Lemma 2.2(4) the surface $\hat{F}_a$ is separating, cutting $M(r_a)$ into a black region and a white region. Two Scharlemann cycles of Γ_b have the same color if the disks they bound lie in the same region.

LEMMA 13.2. *Suppose $n_a = 2$ and $n \geq 1$.*

(1) If $e_1 \cup e_2$ and $e_1' \cup e_2'$ are two Scharlemann bigons of Γ_b of the same color, then either (i) up to relabeling e_i is parallel to e_i' on Γ_a for $i = 1, 2$, or (ii) Γ_a is kleinian, and the four edges e_1, e_2, e_1', e_2' are mutually non-parallel on Γ_a.

(2) If Γ_b has four parallel positive edges then Γ_a is kleinian.

PROOF. (1) If the four edges are in two families of Γ_a then (i) holds. If they are in three families, i.e., e_1 is parallel to e_1' but e_2 is not parallel to e_2', then the nontrivial loop $e_2 \cup e_2'$ on $\hat{F}_a$ is homotopic in $M(r_a)$ to the trivial loop $e_1 \cup e_1'$, which contradicts the incompressibility of $\hat{F}_a$.

Now assume that they are mutually non-parallel. Let G be the subgraph of Γ_a consisting of these four edges and the two vertices of Γ_a. Let B be the side of $\hat{F}_a$ which contains the two Scharlemann bigons. Shrinking the Dehn filling solid torus V_a to its core K_a and cutting B along the two Scharlemann bigons, we obtain a manifold whose boundary consists of the two disk faces of G and two copies of the two Scharlemann bigons, which is a sphere, so by the irreducibility of $M(r_a)$ it bounds a 3-ball. It follows that B is a twisted I-bundle over a Klein bottle K, and K intersects K_a at a single point. Therefore by Lemma 2.12 Γ_a is kleinian.

(2) If Γ_b has four parallel positive edges then they form two Scharlemann bigons of the same color. By Lemma 2.2(2) no two of these edges are parallel on Γ_a, hence by (1) Γ_a is kleinian. □

In the remainder of this section we assume that Γ_a, Γ_b are non-positive, $n_a = 2$, $n = n_b > 4$, and $\max(w_1 + w_2, w_3 + w_4) = 2n - 2$. We may assume without loss of generality that $w_3 + w_4 \leq w_1 + w_2 = 2n - 2$. Since $w_1 + ... + w_4 + 2w_5 = \Delta n \geq 4n$, we have $w_5 = w_6 \geq 2$. Let $\alpha_1 \cup \alpha_2$ (resp. $\beta_1 \cup \beta_2$) be the two edges of $\hat{e}_5$ (resp. $\hat{e}_6$) adjacent to $\hat{e}_1 \cup \hat{e}_2$. Note that these are Scharlemann bigons, hence $\hat{F}_b$ is separating, and n is even. Without loss of generality we may assume that $\alpha_1 \cup \alpha_2$ is a (12)-Scharlemann bigon. Assume that $\beta_1 \cup \beta_2$ is a $(k, k+1)$-Scharlemann bigon.

LEMMA 13.3. *(1) k is even if and only if $w_2 = n - 1$.*
(2) $\{1, 2\} \cap \{k, k+1\} = \emptyset$.

PROOF. (1) This follows from the parity rule. Orient u_1 counterclockwise and u_2 clockwise in Figure 13.1. If $w_2 = n - 1$ then the first edge of $\hat{e}_2$ has label 2 at u_1 and $k + 2$ at u_2, so by the parity rule k must be even. Similarly if $w_2 = n$ or $n - 2$ then k is odd.

(2) If $k = 1$ then by (1) we have $w_2 = n$ or $n - 2$. In the first case the edges of $\hat{e}_1$ would be co-loops, while in the second case the edges of $\hat{e}_2$ would be co-loops.

If $k = 2$ then $w_1 = w_2 = n - 1$ and the edges of $\hat{e}_1$ are co-loops. Similarly if $k = n$ then the edges of $\hat{e}_2$ are co-loops. Since $n - 2 > 2$, all cases contradict Lemma 2.14(2) because the above would imply that there are at least three parallel co-loop edges. □

LEMMA 13.4. *Suppose $w_1 = w_2 = n - 1$. Then for $i = 1, 2$, the edges of $\hat{e}_i$ on Γ_b form a cycle C_i and a chain C_i' disjoint from C_i. Moreover, the vertices of C_1 (C_2') are the set of v_j with j odd, while the vertices of C_2 (C_1') are the set of v_j with j even. The cycles C_1, C_2 are essential on $\hat{\Gamma}_b$.*

PROOF. Let φ_i be the transition function of $\hat{e}_i$. Let h be the number of orbits of φ_i. Since $\hat{e}_i$ has $n - 1$ edges, all but one component of the subgraph of Γ_b consisting of edges of $\hat{e}_i$ are cycles. Therefore $h - 1 \leq 2$ by Lemma 2.14(2). Note also that each orbit contains the same number (n/h) of vertices. Since Γ_a has a Scharlemann bigon, $\hat{F}_b$ is separating and the number of positive vertices of Γ_b is the same as that of negative vertices, hence the number of orbits h is even, so we must have $h = 2$. Hence $\hat{e}_i$ forms exactly one cycle component C_i and one non-cycle component C_i' on Γ_b. Since each odd number appears twice at the endpoints of $\hat{e}_1$, C_1 contains v_j with j odd, and C_1' contains those with j even. For the same reason the edges of $\hat{e}_2$ form a cycle C_2 and a chain C_2'. Since $n/2$ edges of $\hat{e}_2$ have even labels at u_1, C_2 must contain v_j with j even, while C_2' contains v_j with j odd. It follows that $C_1 \cap C_2 = \emptyset$. □

When $w_1 = n - 2$ and $w_2 = n$, the edges of $\hat{e}_2$ form exactly two cycles C_1 and C_2 on Γ_b, essential on $\hat{F}_b$, where the vertices of C_1 (C_2) are the v_j with j odd (even). This is because by Lemma 2.14(2) they cannot form more than two cycles, while Γ_b being non-positive implies that $\hat{e}_2$ cannot form only one cycle. When $w_1 = w_2 = n - 1$, let C_1, C_2 be the cycles given in Lemma 13.4. In either case, let A_1, A_2 be the annuli obtained by cutting $\hat{F}_b$ along $C_1 \cup C_2$. Consider the cycles $\alpha = \alpha_1 \cup \alpha_2$ and $\beta = \beta_1 \cup \beta_2$ on Γ_b. Note that either α and β are in different A_i, or each of them has exactly one edge in each A_i. We say that α, β are *transverse* to C_i in the second case.

LEMMA 13.5. *The cycles α, β are disjoint, and transverse to C_i.*

PROOF. The first statement follows from Lemma 13.3(2), so we need only show that α, β are transverse to C_i.

First assume $\Delta = 5$. Then $w_5 = \frac{1}{2}(\Delta n - (w_1 + ... + w_4)) \geq \frac{n}{2} + 2$. By Lemma 2.3(3) we also have $w_5 \leq \frac{n}{2} + 2$, hence $w_5 = \frac{n}{2} + 2$, in which case the two outermost bigons of the family $\hat{e}_5$ are Scharlemann bigons, with label pair (12) and $(r+1, r+2)$, respectively, where $r = n/2$. By Lemma 2.3(4) the label pair of $\beta_1 \cup \beta_2$ must be either $(1, 2)$ or $(r+1, r+2)$, and by Lemma 13.3 it cannot be the former. Therefore it must be $(r+1, r+2)$.

If α is not transverse to C_i, then it is an essential cycle in one of the annuli, say A_1, obtained by cutting $\hat{F}_b$ along $C_1 \cup C_2$, so β must be an essential cycle in the other annulus A_2. The two cycles α and β separate the vertices of C_1 from C_2, except v_1, v_2, v_{r+1} and v_{r+2} which lie on $\alpha \cup \beta$. On the other hand, the edge e in $\hat{e}_5$ adjacent to α_2 has label pair $(3, n)$, so there is an edge on Γ_b connecting v_3 to v_n. Since n is even, the vertices v_3, v_n belong to different C_i, but since $n > 4$, neither 3 nor n belongs to the set $\{1, 2, r+1, r+2\}$, which is a contradiction.

Now assume $\Delta = 4$. In this case the jumping number $J(r_a, r_b) = \pm 1$. Consider the two negative edges e', e'' of Γ_a with label 2 at u_1. Note that their endpoints at u_1 are separated by the label 2 endpoints of α_1, α_2, hence by the Jumping Lemma, on Γ_b the endpoints of e', e'' at v_2 are separated by those of α; in other words, e', e'' are on different sides of the cycle α. Assume that $v_2 \in C_2$ is positive. If α is not transverse to C_2 then all positive edges at v_2 must be on one side of α because the other side is shielded by the cycle C_1, which contains only negative vertices. This is a contradiction. Therefore α, and hence β, must be transverse to C_i. □

LEMMA 13.6. *Each edge of $\hat{e}_1 \cup \hat{e}_2$ is either on $C_1 \cup C_2$ or parallel to an edge of $C_1 \cup C_2$ on Γ_b.*

PROOF. Let C_1, C_2 and α, β be as above. By definition C_2 consists of the edges in $\hat{e}_2$ with even labels. Let C'_1 be the edges of $\hat{e}_1$ with even labels. Because of symmetry it suffices to show that each edge of C'_1 is parallel to an edge in C_2.

Note that $\alpha \cap C_2 = v_2$. Let $v_t = \beta \cap C_2$. (Thus t is the even label of the Scharlemann bigon $\beta_1 \cup \beta_2$ in Γ_a.) Since $w_1 + w_2 = 2n - 2$ and the edges adjacent to $\hat{e}_1 \cup \hat{e}_2$ on Γ_a are the (12)-Scharlemann bigon $\alpha_1 \cup \alpha_2$ and the $(t, t+1)$- or $(t-1, t)$-Scharlemann bigon $\beta_1 \cup \beta_2$, we see that $2, t$ are the only even labels appear three times among the endpoints of edges in $\hat{e}_1 \cup \hat{e}_2$, hence on Γ_b the edges of C'_1 form a chain with endpoints at v_2, v_t, and possibly some cycle components. Therefore $C'_1 - v_2 \cup v_t$ is disjoint from $\alpha \cup \beta \cup C_1$, hence lies in the interior of the two disks obtained by cutting $\hat{F}_b$ along $\alpha \cup \beta \cup C_1$. By Lemma 2.14(1) this implies that C'_1 has no cycle component, and hence is a chain. Since C'_1 contains all vertices of C_2, this also implies that one component of $C_2 - v_2 \cup v_t$ contains no vertices of Γ_b; in other words, the two vertices v_2, v_t are adjacent on C_2.

Let q, p be the transition number of $\hat{e}_1, \hat{e}_2$, respectively. Since C_2 has an edge connecting v_2 to v_t, we have $p \equiv \pm(t - 2) \mod n$. Since C'_1 is a chain of length $(n/2) - 1$ connecting v_2, v_t, we have $((n/2) - 1)q \equiv \pm(t - 2) \mod n$. An edge of C'_1 has even labels on both endpoints, so q is even, hence $((n/2) - 1)q \equiv -q \mod n$. It follows that $p \equiv \pm q \mod n$, which implies that each edge e' of C'_1 has its endpoints on adjacent vertices of C_2. Let e be the edge of C_2 connecting these two vertices. Since e' has interior disjoint from $\alpha \cup \beta \cup C_1$, it must be parallel to e. □

PROPOSITION 13.7. *The case that Γ_a, Γ_b are non-positive, $n_a = 2$, $n = n_b > 4$, and $w_3 + w_4 \leq w_1 + w_2 = 2n - 2$, is impossible.*

PROOF. First assume that $w_1 = n - 2$ and $w_2 = n$. By Lemma 13.3, the label pair of β is $(k, k+1)$, where k is odd and $k \neq 1$. If $k = n - 1$ then the edges of $\hat{e}_2$ are co-loops, which contradicts Lemma 2.14(2). Therefore $n - 3 \geq k \geq 3$.

Since the label sequence of $\hat{e}_1$ at u_2 is $k+2, ..., n, 1, ..., k-1$, the above implies that there are adjacent edges $e'_1, e'_n \in \hat{e}_1$ with labels 1 and n at u_2, respectively. By Lemma 13.6 each edge of $\hat{e}_1$ is parallel to some edge of $\hat{e}_2$ on Γ_b, hence the transition function ψ_1 of $\hat{e}_1$ is either equal to ψ_2 of $\hat{e}_2$, or ψ_2^{-1}, but since the two edges of $\hat{e}_1 \cup \hat{e}_2$ with label n at u_1 have labels $k+1$ and $k-1$ respectively at u_2, the first case is impossible, hence $\psi_1 = \psi_2^{-1}$. Let $\hat{e}_2 = e_1 \cup ... \cup e_n$, where e_i has label i at u_1. Since e_1 is the only edge of $\hat{e}_2$ with label 1 at u_1, it must be the one that is parallel to e'_1 on Γ_b. Similarly, e_n is parallel to e'_n on Γ_b. This is a contradiction to Lemma 2.19.

Now assume that $w_1 = w_2 = n - 1$. As above, let $\hat{e}_2 = e_2 \cup ... \cup e_n$, where e_i has label i at u_1. The label sequence of $\hat{e}_1$ at u_2 is $k+1, k+2, ..., n, 1, ..., k-1$. By

Lemma 13.3, k is even, and $\{1,2\} \cap \{k, k+1\} = \emptyset$, so $n - 2 \geq k \geq 4$. It follows that there are three consecutive edges e'_n, e'_1, e'_2 of $\hat{e}_1$ such that e'_i has label i at u_2. For the same reason as above, e'_2 is parallel to e_2 and e'_n is parallel to e_n on Γ_b. Since the number of edges between e'_n and e'_2 is 1 while the number of edges between e_2 and e_n is $n - 3 > 1$ on Γ_a, this is a contradiction to Lemma 2.19. □

14. The case $n_a = 2$, $n_b > 4$, Γ_1, Γ_2 non-positive, and $w_1 = w_2 = n_b$

In this section we consider the case that $n_a = 2$, $n = n_b > 4$, Γ_1, Γ_2 non-positive, and $w_1 = w_2 = n$. We will also assume without loss of generality that $w_3 \geq w_4$. Let $\hat{e}_1 = e_1 \cup ... \cup e_n$, $\hat{e}_2 = e'_1 \cup ... \cup e'_n$, and assume that e_i, e'_i have label i at u_1.

Let r be such that the label of the endpoint of e_1 on ∂u_2 is $r + 1$. One can check that both e_i, e'_i have label $r + i$ at ∂u_2.

Since Γ_b is non-positive, the vertices of Γ_b cannot all be parallel, so the edges of $\hat{e}_1$ form at least two cycles on Γ_b. By Lemma 2.14(2) they form exactly two cycles $C_1 \cup C_2$ on Γ_b.

LEMMA 14.1. *Γ_a is not kleinian. In particular, Γ_b cannot contain four parallel positive edges.*

PROOF. If Γ_a is kleinian then by Lemma 6.2(4) there is a free orientation reversing involution ϕ of $(\hat{F}_a, \Gamma_a)$, which maps u_1 to u_2, and is label preserving. If there is no loop on Γ_a (i.e. $w_5 = w_6 = 0$), then $\Delta = 4$ and $w_i = n$ for all i, so the label sequences of $\hat{e}_i$ at u_1 are all the same. The above implies that the label sequences of $\hat{e}_i$ at u_2 are also the same as those at u_1, so the transition function φ defined by $\hat{e}_i$ is the identity map and hence all edges of Γ_a are co-loops, contradicting the 3-Cycle Lemma 2.14(2).

Now assume $w_5 = w_6 > 0$. Then ϕ maps $\hat{e}_1 \cup \hat{e}_2$ to either $\hat{e}_1 \cup \hat{e}_2$ or $\hat{e}_3 \cup \hat{e}_4$. In the first case since ϕ is label preserving and orientation reversing on the torus, the label sequence of $\hat{e}_1$ at u_2 is the same as that of $\hat{e}_1$ at u_1, hence all edges of $\hat{e}_1$ are co-loops and we have a contradiction to Lemma 2.14(2). In the second case $w_3 = w_4 = w_1 = w_2 = n$, so $\Delta = 5$ and $w_5 = w_6 = n/2$. We have assumed that $\hat{e}_1$ has label sequence $1, 2, ..., n$ at u_1, so $\hat{e}_3$ has the same label sequence at u_2. Since $w_5 = n/2$, the label sequence of $\hat{e}_1$ at u_2 is $k+1, k+2, ...n, 1, ..., k$, where $k = n/2$. Therefore ϕ is of period 2, so it has $n/2 > 2$ orbits, which again contradicts Lemma 2.14(2).

The second statement follows from the above and Lemma 13.2(2). □

LEMMA 14.2. *The edges e_i, e'_i are parallel on Γ_b.*

PROOF. The cycles $C_1 \cup C_2$ defined at the beginning of the section cut the torus $\hat{F}_b$ into two annuli A_1, A_2. Each e'_i lies in one of the A_j and has the same endpoints as e_i, so if it is not parallel to e_i and $e_i \subset C_1$ then it is parallel to $C_1 - e_i$. There are at most two such e'_i for C_1, one in each A_j. Since $n > 4$, C_1 contains at least three edges, hence there exists some e'_j parallel to $e_j \subset C_1$.

Assume e_i is not parallel to e'_i, and let e_j, e'_j be parallel on $\hat{F}_b$, which exist by the above. Let D (resp. D') be the disk on F_a that realizes the parallelism between e_i, e_j (resp. e'_i, e'_j), and let D'' be the disk between e_j and e'_j on F_b. Shrinking V_b to its core K_b, $B = D \cup D' \cup D''$ becomes a disk in $M(r_b)$ with $\partial B = e_i \cup e'_i$, which contradicts the fact that $\hat{F}_b$ is incompressible in $M(r_b)$. □

14. THE CASE $n_a = 2$, $n_b > 4$, Γ_1, Γ_2 NON-POSITIVE, II

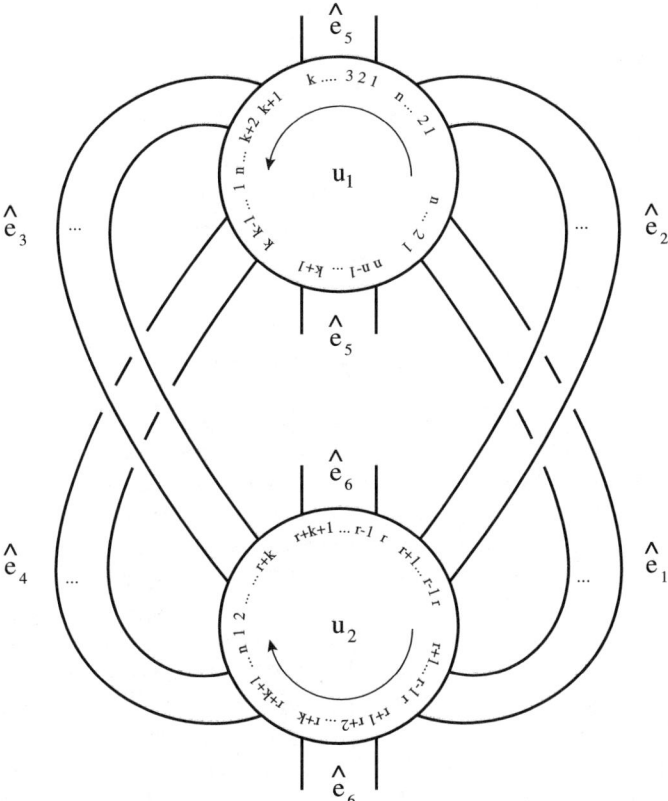

Figure 14.1

LEMMA 14.3. $w_3 + w_4 \neq n$.

PROOF. Assume to the contrary that $w_3 + w_4 = n$. We have $\Delta = 4$ as otherwise there would be n parallel positive edges in $\hat{e}_5$, contradicting Lemma 2.3(3). Now $w_5 = w_6 = n/2$, so the graph Γ_a is as shown in Figure 14.1, where $k = n/2$. Since $\Delta = 4$, we may assume that the jumping number is 1.

Let i be a label such that $1 \leq i \leq k$, so it appears on the top of the vertex u_1 in Figure 14.1. Consider the vertex v_i of Γ_b, see Figure 14.2. By Lemma 14.2 e_i of $\hat{e}_1$ is parallel to e'_i of $\hat{e}_2$. Since e_i and e'_i have the same label 1 at v_i, there is an edge of $\hat{e}_3 \cup \hat{e}_4$ between them. Similarly there are parallel edges e_j, e'_j with label 2 at v_i, and there is another edge between them. See Figure 14.2. From the labeling we see that the two negative edges at v_i (corresponding to loops in Γ_a) must be adjacent to each other.

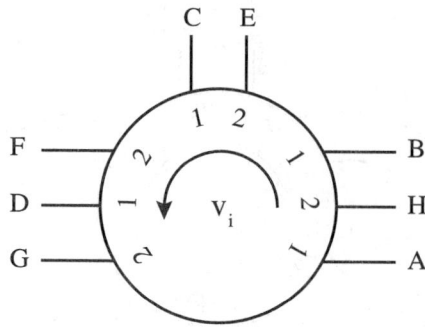

Figure 14.2

On ∂u_1 the i-labels appear on endpoints of edges in the order of A, B, C, D, where $A \in e_i$, $B \in e'_i$, C is a loop in $\hat{e}_5$, and $D \in \hat{e}_3 \cup \hat{e}_4$. Since the jumping number is 1, the 1-labels at v_i also appear in the same order. In Figure 14.2 this implies that the negative edge C appears on the top of the vertex.

Now consider the four edges labeled 2 at v_i, denoted by E, F, G, H, where E is the negative edge, which is uniquely determined. Since F, G are parallel positive edges on Γ_b, they are the e_j, e'_j given above, belonging to $\hat{e}_1 \cup \hat{e}_2$. On ∂u_2 this implies that the endpoint of the loop E labeled i appears on the top of ∂u_2 in Figure 14.1. Since i is any label between 1 and $n/2$, it follows that the labels on the top of ∂u_2 must be $1, 2, ..., k$, so the integer r in the figure satisfies $r = k$. However, in this case the edges of $\hat{e}_1$ would form cycles of length 2 in Γ_b, which is a contradiction to Lemma 2.14(2). □

LEMMA 14.4. $w_3 = n$, and $0 < w_4 < n$. Moreover, an edge e'' of $\hat{e}_3$ with label j at u_2 is parallel to the edges e_j and e'_j.

PROOF. We have assumed $w_4 \leq w_3 \leq n$. If $w_4 = n$ then the argument of Lemma 14.2 applied to $\hat{e}_3, \hat{e}_4$ shows that each edge of $\hat{e}_3$ is parallel to exactly one edge of $\hat{e}_4$. On the other hand, since the two parallel edges e_i, e'_i in the proof of Lemma 14.2 have the same label 1 at the vertex v_i, there must be another edge e''_i in $\hat{e}_3 \cup \hat{e}_4$ between e_i and e'_i. Together with the other edge in $\hat{e}_3 \cup \hat{e}_4$ which is parallel to e''_i, we get four parallel positive edges in Γ_b, which contradicts Lemma 14.1. Therefore $w_4 < n$.

Recall from Lemma 14.2 that the edges e_i and e'_i are parallel in Γ_b, with the same label 1 at u_i, so there must be another edge $e''_i \in \hat{e}_3 \cup \hat{e}_4$ between them. Note also that if e_i has label $i + r$ at u_2 then e''_i has the property that it has label i at u_2 and $i + r$ at u_1. This is true for all i, so either $\hat{e}_3$ and $\hat{e}_4$ have the same transition function, or these e''_i all belong to the same family. The first case happens only if $w_3 + w_4 \equiv 0 \mod n$, which is impossible because by Lemma 14.3 we have $w_3 + w_4 \neq n$, while $w_3 \leq n$ and by the above we have $w_4 < n$. Therefore all the e''_i belong to $\hat{e}_3$. Since $w_3 \leq n$, this implies that $w_3 = n$. Again by Lemma 14.3 we have $w_4 \neq 0$. □

14. THE CASE $n_a = 2$, $n_b > 4$, Γ_1, Γ_2 NON-POSITIVE, II

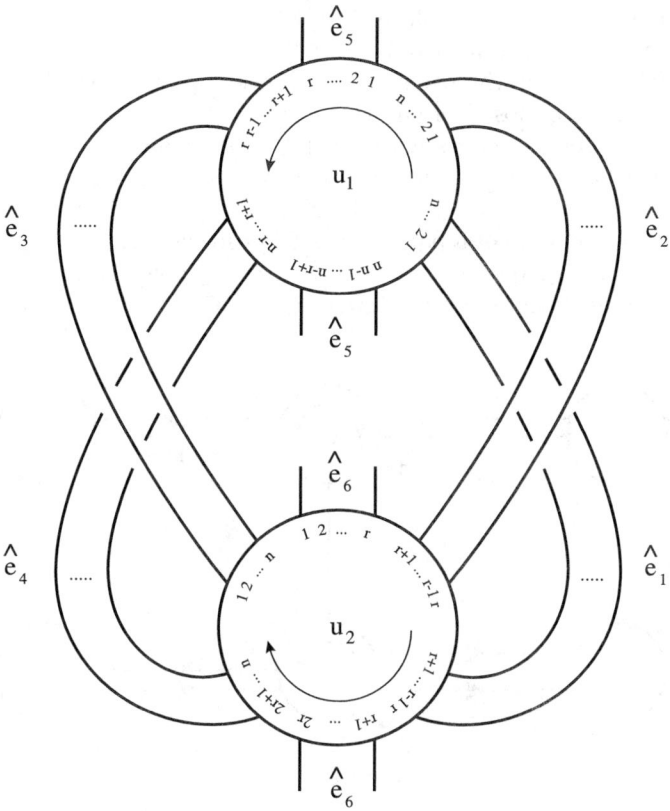

Figure 14.3

LEMMA 14.5. *The label sequence of $\hat{e}_3$ is $1, 2, ..., n$ at u_2, and $1+r, 2+r, ..., n, 1, ..., r$ at u_1. The labels of Γ_a are as shown in Figure 14.3.*

PROOF. First assume that the label sequence of $\hat{e}_3$ at u_2 is not $1, 2, ..., n$. Then there is a pair of adjacent parallel edges e''_n, e''_1 with label n and 1 at u_2, respectively. By Lemma 14.4 $e_1 \cup e''_1$ and $e_n \cup e''_n$ are parallel pairs on Γ_b. Since e''_1, e''_n are adjacent on Γ_a while e_1, e_n are not, this is a contradiction to Lemma 2.19. Therefore the label sequence of $\hat{e}_3$ at u_2 must be $1, 2, ..., n$.

Since by Lemma 14.4 the edge e''_i connects v_i to v_{i+r} with label 2 at v_i and 1 at v_{i+r}, we see that on Γ_a it has label i at u_2 and $i+r$ at u_1, hence the label sequence of $\hat{e}_3$ at u_1 is $r+1, ..., n, 1, ..., r$. The labels of $\hat{e}_1, \hat{e}_2, \hat{e}_3$ determine those of the loops, and hence those of $\hat{e}_4$. Therefore Γ_a must be as shown in Figure 14.3. □

LEMMA 14.6. *(1) The jumping number $J = \pm 1$.*
(2) Orient the negative edges of Γ_a from u_1 to u_2. Then on Γ_b the edges of $\hat{e}_1$ form two essential cycles of opposite orientation on $\hat{F}_b$.

PROOF. (1) Since $\Delta = 4$ or 5, the jumping number is either ± 1 or ± 2. Let e_i, e'_i be the edges of $\hat{e}_1, \hat{e}_2$, respectively, with label i at u_1. If $J = \pm 2$ then these edges are not adjacent among the 1-edges at v_i in Γ_b. Since by Lemma 14.2 they are parallel in Γ_b, there would be more than $2n_a = 4$ parallel edges in Γ_b, which contradicts Lemma 2.2(2). Therefore $J = \pm 1$. Changing the orientation of $\hat{F}_b$ if necessary, we may assume that $J = 1$.

(2) Now let C_1, C_2 be the cycles on Γ_b consisting of edges of $\hat{e}_1$. We need to show that they are of opposite orientation.

Let $a_1, a_2, \bar{a}, a_3$ be the edges with label 1 at u_1, where $a_i \in \hat{e}_i$ for $i = 1, 2, 3$, and $\bar{a} \in \hat{e}_5$. Note that they appear in this order on ∂u_1. Since $J = 1$, they also appear in this order on ∂v_1 in Γ_b, see Figure 14.4. By the proof of Lemma 14.3 we see that a_3 is in the middle of a pair of parallel positive edges incident to v_1, which is not parallel to a_1, a_2, hence the orientation of C_1 must be as shown in Figure 14.4, where C_1 is represented by the lower level chain.

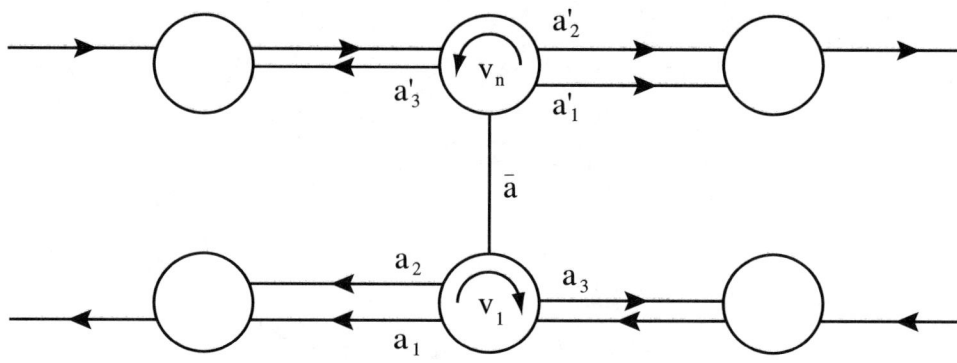

Figure 14.4

Now consider the edges labeled n at u_1. There are 5 of them if $\Delta = 5$, but we only consider $\bar{a}$ and the edges a_1', a_2', a_3', where $a_i' \in \hat{e}_i$. The order of the label n endpoints of these edges on ∂u_1 is $a_3', \bar{a}, a_1', a_2'$, while the orientation of v_n is opposite to that of v_1. Therefore these edges appear on ∂v_n as shown in Figure 14.4. We see that C_1, C_2 are of opposite orientation on $\hat{F}_b$. □

PROPOSITION 14.7. *Suppose $n_a = 2$, $n > 4$, Γ_a, Γ_b are non-positive, and $w_1 = w_2 = n$.*

(1) On Γ_b each edge of $\hat{e}_4$ connects a pair of adjacent vertices of some C_i, but is not parallel to an edge of C_i.

(2) Two adjacent edges of $\hat{e}_4$ lie in different annuli of $\hat{F}_b - \cup C_i$.

(3) $w_4 = w_5 = w_6 = 2$, $\Delta = 4$, and $n = 6$.

(4) The graphs Γ_a, Γ_b and their edge correspondence are as shown in Figure 14.5, where e_i (resp. e_i') is the edge in $\hat{e}_1$ (resp. $\hat{e}_2$) with label i at u_1, and the edge between e_i, e_i' is the edge of $\hat{e}_3$ with label i at u_2.

PROOF. (1) From Figure 14.3 we see that an edge e of $\hat{e}_4$ with label i at u_1 has label $i + r$ at u_2. Since the transition function of $\hat{e}_1$ also maps i to $i + r$, v_i and v_{i+r} are connected by the edge e_i of $\hat{e}_1$, and hence are adjacent on one of the cycles C_j. This proves the first part of (1). By Lemmas 14.2 and 14.4 each edge e_i of C_j is parallel to an edge e_i' in $\hat{e}_2$ and an edge e_i'' in $\hat{e}_3$, so if e is parallel to e_i then there would be four parallel positive edges in Γ_b, which would contradict Lemma 14.1.

(2) By Lemma 14.6(2) the two cycles C_1, C_2 have opposite orientations. Without loss of generality we may assume that the orientations of C_i are as shown in Figure 14.5. Recall that C_1 is the cycle containing the vertex v_1. By Lemma 14.6(1) we may assume without loss of generality that the jumping number of the graphs is 1. Let e be an edge of $\hat{e}_4$ with label k at u_1, and let e_k, e'_k, e''_k be the edges of $\hat{e}_1, \hat{e}_2, \hat{e}_3$ with label k at u_1. Then the endpoints of these edges appear at ∂u_1 in the order e_k, e'_k, e''_k, e, so on ∂v_k they appear in the same order. If v_k is in C_1 then the orientation of C_1 points to the left and the orientation of v_k is clockwise, so e is in the annulus below C_1. If v_k is in C_2 then the orientation of C_2 points to the right and the orientation of v_k is counterclockwise, so again e is in the annulus below C_2. Since the labels of adjacent edges of $\hat{e}_4$ belong to different C_i in Γ_b, it follows from the above that they are in different annuli of $\hat{F}_b - \cup C_i$.

(3) Since each C_i contains $n/2 > 2$ vertices, there cannot be two edges on the same side of C_i connecting two different pairs of adjacent vertices and yet not parallel to an edge of C_i. Hence by (1) and (2) $\hat{e}_4$ contains at most two edges. By Lemma 14.4 $w_4 > 0$, and from the labeling in Figure 14.3 we see that w_4 is even. Therefore $w_4 = 2$.

If $\Delta = 5$ then the loop family of Γ_a at u_1 contains $n-1$ edges. This contradicts Lemma 2.3(3) for $n > 4$. Hence $\Delta = 4$.

Let e be an edge of $\hat{e}_4$ with endpoints on v_i and v_j in C_2, lying on the annulus A below C_2. By (1) it is not parallel to the edge on C_2 connecting v_i, v_j, so on A it separates C_1 from other vertices of C_2, hence there is no edge in A connecting C_1 to vertices of C_2 except possibly v_i and v_j. By Lemmas 14.2 and 14.4 there are three parallel edges for each edge of C_i. Together with e, they contribute 7 edge endpoints to each of v_i and v_j, therefore $\Delta = 4$ implies that there are at most two edges in A connecting C_1 to C_2, one for each of v_i, v_j. Note that these correspond to loop edges in Γ_a. Therefore the two annuli give rise to at most 4 loops in Γ_a, so $w_5 = w_6 \leq 2$. Since $n > 4$ and $2w_5 + w_4 = (\Delta - 3)n = n$, it follows that $n = 6$, and $w_5 = w_6 = 2$.

(4) By Lemma 14.5 Γ_a is the graph in Figure 14.3. We have $w_4 = w_5 = w_6 = 2$ and $w_1 = w_2 = w_3 = n = 6$, hence Γ_a is as shown in Figure 14.5(a).

The edges in $\hat{e}_2, \hat{e}_3$ are parallel to those in $\hat{e}_1$, as shown in Lemmas 14.2 and 14.5, therefore they form families of three parallel edges, as shown in Figure 14.5(b). Orientations are from u_1 to u_2 on Γ_a, so the tails of these edges are labeled 1 and the heads labeled 2 on Γ_b. The two edges in $\hat{e}_4$ connect v_4, v_6 and v_3, v_5 respectively, and by (1) and (2) they are not parallel to edges in C_i and lie in different annuli of $\hat{F}_b - C_1 \cup C_2$, hence we may assume that they look like that in Figure 14.5(b). The four edges in $\hat{e}_5$ and $\hat{e}_6$ are now determined by the labeling of the edges and the vertices on Γ_b. The labeling of the weight 3 families in Γ_b are determined by the single edges and the assumption that the jumping number is 1. $\square$

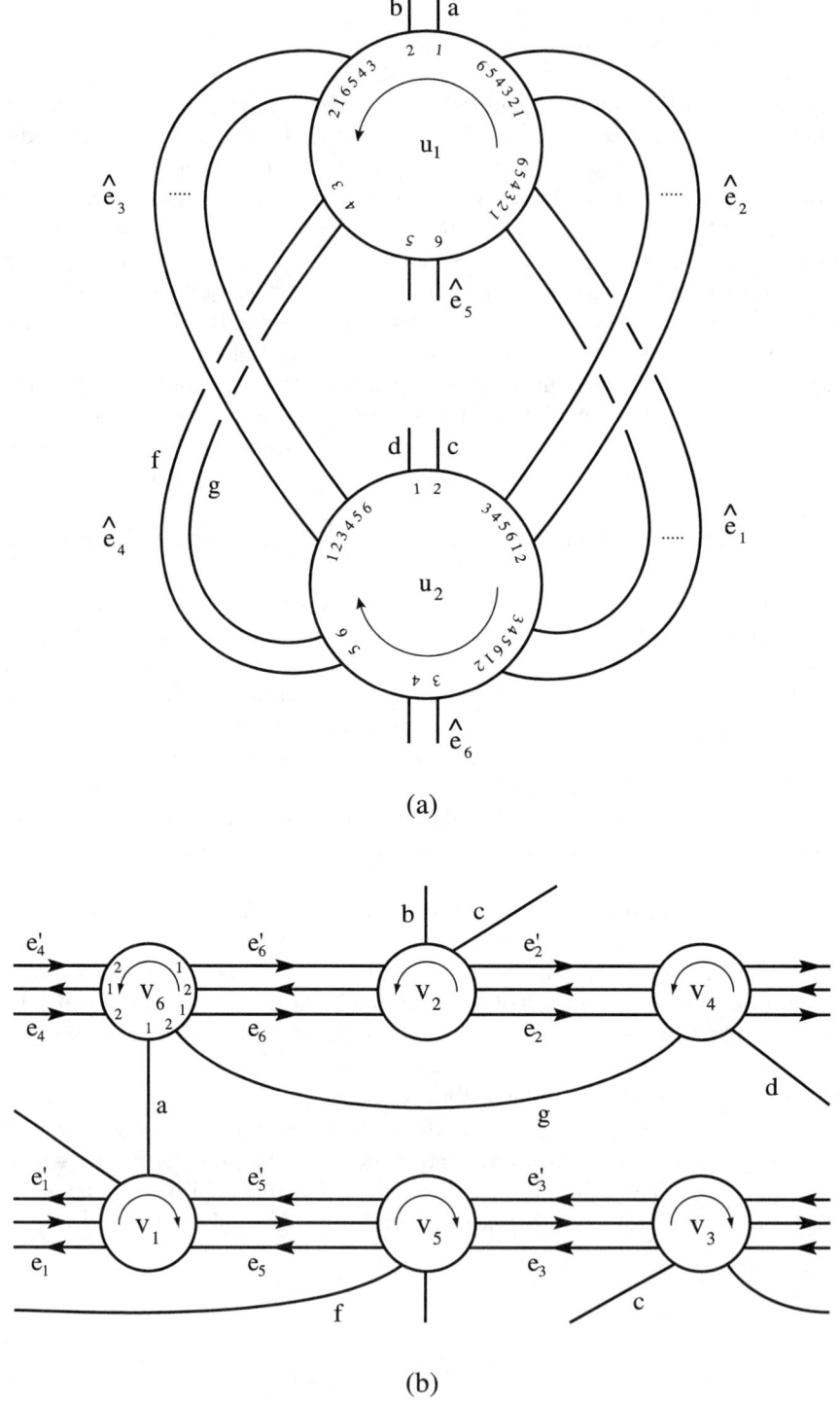

Figure 14.5

15. Γ_a with $n_a \leq 2$

The next few sections deal with the case that $n_a \leq 2$ and $n_b \leq 4$. In this section we set up notation and give some preliminary results.

We use $G = (b_1, b_2, b_3)$ to denote a graph G on a torus with one vertex and three families of edges weighted b_1, b_2, b_3. Similarly, denote by $G = (\rho; a_1, ..., a_4)$ a graph G on a torus which has two vertices, two families of loops of weight ρ, and four families of edges $\hat{e}_i$ with weight sequence $a_1, ..., a_4$ around the vertices. It is possible that ρ and some of the a_i may be zero. When $\rho = 0$ we will simply write $G = (a_1, ..., a_4)$. Note that the weight sequence is defined up to cyclic rotation and reversal of order. When $\rho = 0$, any weight 0 can be moved around without changing the graph, hence $(2, 2, 0, 0)$ is equivalent to $(2, 0, 2, 0)$, but $(1; 2, 0, 2, 0)$ is different from $(1; 2, 2, 0, 0)$ and $(3, 1, 3, 1)$ is different from $(3, 3, 1, 1)$. When it is necessary to indicate whether the vertices of G are parallel or antiparallel, we write $G = +(\rho; a_1, ..., a_4)$ if the vertices of G are parallel, and $G = -(\rho; a_1, ..., a_4)$ otherwise.

Suppose $n_a = 2$ and n_b is even. Then Γ_a is of the form $(\rho; a_1, ..., a_4)$. Let e be an edge of $\hat{e}_i$. Define $\epsilon_i = 0$ if the two labels at ∂e have the same parity, and $\epsilon_i = 1$ otherwise. The assumption that n_b is even implies that this is independent of the choice of $e \in \hat{e}_i$. If $n_b = 2$ then $\epsilon_i = 0$ if and only if one (and hence all) of the edges in $\hat{e}_i$ is a co-loop edge, in which case we say that $\hat{e}_i$ is co-loop.

LEMMA 15.1. **(The Congruence Lemma.)** *Suppose $n_a = 2$, and n_b is even. Let $\hat{e}_i, \hat{e}_j$ be non-loop edges in $\hat{\Gamma}_a$. Let a_k be the weight of $\hat{e}_k$.*

(1) If $\hat{\Gamma}_a$ has no loops and $a_i, a_j \neq 0$ then $a_i + \epsilon_i \equiv a_j + \epsilon_j \mod 2$. In particular, if $n_b = 2$ then $a_i \equiv a_j \mod 2$ if and only if $\hat{e}_i$ and $\hat{e}_j$ are both co-loop or both non co-loop.

(2) Suppose $\hat{\Gamma}_a$ has loops, and Γ_b has the property that v_i is parallel to v_j if and only if i, j have the same parity. If either $a_i, a_j \neq 0$ or the endpoints of $\hat{e}_i, \hat{e}_j$ at u_1 are on the same side of the loop at u_1, then $a_i \equiv a_j \mod 2$.

PROOF. (1) Delete edges of $\hat{\Gamma}_a$ with zero weight. We need only prove the statement for adjacent edges $\hat{e}_1, \hat{e}_2$ of $\hat{\Gamma}_a$ with non-zero weight. Let $e_1, ..., e_{a_1}$ and $e'_1, ..., e'_{a_2}$ be the edges in $\hat{e}_1$ and $\hat{e}_2$, respectively, so that e'_1 is adjacent to e_{a_1} on ∂u_1. Then e'_{a_2} is adjacent to e_1 on ∂u_2. Without loss of generality assume that the label of e_i at u_1 is i. Let t_j be the transition number of $\hat{e}_j$ from u_1 to u_2. Then the label of e_1 at u_2 is $1 + t_1$. On the other hand, the label of e'_1 at u_1 is $a_1 + 1$, so the label of e'_{a_2} at u_1 is $a_1 + a_2$, and the label of e'_{a_2} at u_2 is $a_1 + a_2 + t_2$. See Figure 15.1. Since e'_{a_2} is adjacent to e_1 on u_2, the label of e_1 on u_2 is $a_1 + a_2 + t_2 + 1$. By definition $\epsilon_j \equiv t_j \mod 2$, therefore these two equations give

$$1 + \epsilon_1 \equiv a_1 + a_2 + \epsilon_2 + 1 \mod 2$$

It follows that $a_1 \equiv a_2$ if and only if $\epsilon_1 \equiv \epsilon_2 \mod 2$.

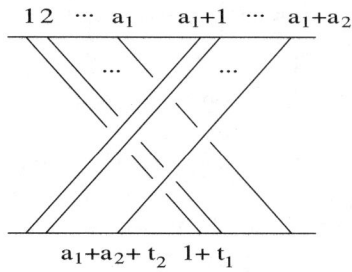

Figure 15.1

(2) We need only prove the statement for adjacent non-loop edges $\hat{e}_1, \hat{e}_2$ of $\hat{\Gamma}_a$. By assumption v_p, v_q are parallel if and only $p \equiv q$ mod 2, so the parity rule gives $\epsilon_1 \equiv \epsilon_2$ mod 2. Also, a loop edge in Γ_a must have labels of different parity at its two endpoints, so if the endpoints of $\hat{e}_1, \hat{e}_2$ at u_1 are on the same side of the loop then $a_1 + a_2 \equiv 0$ mod 2.

It remains to prove the case that the endpoints of $\hat{e}_1, \hat{e}_2$ at u_1 are on different side of the loop at u_1, $a_1, a_2 \neq 0$, and $\hat{e}_1, \hat{e}_2$ are adjacent among nonempty non-loop edges. Note that the endpoints of $\hat{e}_1, \hat{e}_2$ are on different sides of the loop at u_1 if and only if their other endpoints are on different sides of the loop at u_2. Since the number of loops at the two vertices are the same, the distance between the endpoints of e_{a_1} and e'_1 on ∂u_1 is the same as that of e'_{a_2} and e_1 on ∂u_2, hence the argument in case (1) can be modified to show that if $a_1, a_2 \neq 0$ then $a_1 \equiv a_2$ mod 2. More explicitly, if there are k loops between them then the endpoints of e'_{a_2} are labeled $a_1 + a_2 + k$ at u_1, and $a_1 + a_2 + k + t_2$ at u_2, and we have $a_1 + a_2 + k + t_2 + k + 1 = 1 + t_1$, hence the result follows because $t_i \equiv \epsilon_i$ and $\epsilon_1 \equiv \epsilon_2$ mod 2. □

LEMMA 15.2. *Suppose Γ_b is positive, and contains a black bigon $e_1 \cup e_2$ and a white bigon $e'_1 \cup e'_2$. Then on Γ_a the four edges e_1, e_2, e'_1, e'_2 cannot be contained in two families of parallel edges.*

PROOF. Recall that no two edges are parallel on both graphs, so if the lemma is not true then we may assume that e_i is parallel to e'_i on Γ_a. Let B_i be the disk on F_a realizing the parallelism, and let D, D' be the bigon on F_b bounded by $e_1 \cup e_2$ and $e'_1 \cup e'_2$, respectively. Then $A = D \cup D' \cup B_1 \cup B_2$ is either a Möbius band or an annulus. The first case contradicts the fact that a hyperbolic manifold M contains no Möbius bands. In the second case A contains a single white bigon and hence each of its boundary components intersects a curve of slope r_a transversely at a single point. Since e_1 is an essential arc on both F_a and A, A cannot be boundary parallel, and hence is essential in M, which is again a contradiction to the hyperbolicity of M. □

16. The case $n_a = 2$, $n_b = 3$ or 4, and Γ_1, Γ_2 non-positive

Throughout this section we assume that $n_a = 2$, $n_b = 3$ or 4, and both Γ_1, Γ_2 are non-positive. We will show that in this case there are only three possibilities for the pair (Γ_a, Γ_b), given in Figures 16.6, 16.8 and 16.9. The following lemma rules out the possibility that $n_b = 3$.

LEMMA 16.1. *The case $n_a = 2$, $n_b = 3$ and Γ_a, Γ_b non-positive, is impossible.*

16. THE CASE $n_a = 2$, $n_b = 3$ OR 4, AND Γ_1, Γ_2 NON-POSITIVE

PROOF. The graph Γ_a contains at most one loop at each vertex as otherwise it would contain a Scharlemann bigon, which contradicts Lemma 2.2(4) because $n_b = 3$ implies that $\hat{F}_b$ is non-separating. There are at most four families of edges on Γ_a connecting u_1 to u_2, containing a total of at least $\Delta n_b - 2 \geq 10$ edges, hence there is a family containing 3 edges $e_1 \cup e_2 \cup e_3$. These are positive edges in Γ_b, and we may assume that e_i has label i at u_1. Since one of the vertices of Γ_b, say v_1, is anti-parallel to the other two vertices, the edge e_1 is a loop on Γ_b, so its label on u_2 is also 1. Since u_1, u_2 are antiparallel, we see that the label of e_i at u_2 is i for $i = 1, 2, 3$, hence they are all co-loop edges on Γ_a. This is a contradiction to the 3-Cycle Lemma 2.14(2). □

We will assume in the remainder of this section that $n_b = 4$. By Lemma 13.1 the graph $\hat{\Gamma}_a$ is as shown in Figure 13.1. Note that $\hat{e}_1, \hat{e}_2$ are on the same side of the loop at each u_i. Denote by w_i the weight of $\hat{e}_i$, and put $\lambda = w_5 = w_6$. Then we can denote Γ_a by $(\lambda; w_1, w_2, w_3, w_4)$, and by Lemma 13.1(2) we may assume that $w_3 + w_4 \leq w_1 + w_2 = 6$ or 8. By Lemmas 2.3(1) and 2.3(3) we have $\lambda, w_i \leq 4$. Also, counting the number of edges incident to u_i gives

$$\sum_{i=1}^{4} w_i + 2\lambda = 4\Delta$$

LEMMA 16.2. *(1) If $w_i \geq 3$ then $s_i = 2$, where s_i is the transition number of $\hat{e}_i$.*

(2) v_1 is parallel to v_3 and antiparallel to v_2 and v_4.

(3) (w_1, w_2) and (w_3, w_4) cannot be $(3, 2)$, $(3, 3)$ or $(3, 4)$.

PROOF. (1) Let s_i be the transition number of $\hat{e}_i$. By the 3-Cycle Lemma (2.14(2)) we have $s_i \neq 0$. If $s_1 = \pm 1$ then all vertices of Γ_b would be parallel, which is a contradiction to the assumption that Γ_b is non-positive. Since $n_b = 4$, the only remaining possibility is that $s_i = 2$.

(2) If $\lambda \geq 3$ then Γ_a contains a Scharlemann cycle among the loops, so $\hat{F}_b$ is separating and the result follows. If $\lambda \leq 2$ then the equation $\sum w_i + 2\lambda = 4\Delta$ gives $w_i \geq 3$ for some i. By (1) and the parity rule, v_j is parallel to v_{j+2}, hence the result follows because Γ_b is non-positive.

(3) Assume $w_1 = 3$. By the equation above, $\lambda > 0$, hence by Lemma 15.1(2) w_2 is odd. The transition function of $\hat{e}_1$ is given by (1), and it will determine that of $\hat{e}_2$. If $w_2 = 3$ then one can check that the transition function of $\hat{e}_2$ would map j to j, which would be a contradiction to (1). □

LEMMA 16.3. $\lambda \geq 2$.

PROOF. First assume $\lambda = 0$. Then $\Gamma_a = (0; 4, 4, 4, 4)$. By Lemma 16.2(1) all edges of Γ_a have label pair $(1, 3)$ or $(2, 4)$, see Figure 16.1. Thus $\hat{\Gamma}_b$ is a union of two cycles, hence all edges from v_1 to v_3 in Γ_b are equidistant. Since two of these edges are in $\hat{e}_1$ and are not equidistant on Γ_a, this is a contradiction to the Equidistance Lemma 2.17.

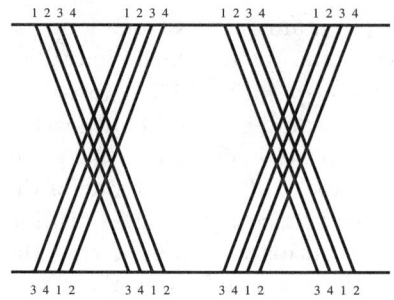

Figure 16.1

If $\lambda = 1$, then $w_i \geq 2$ for $i = 1, ..., 4$, so $w_i \neq 3$ by Lemma 16.2(3). Hence $\Gamma_a = (1; 4, 4, 4, 2)$. One can check that the one of the families of weight 4 would have the same label on the two endpoints of any of its edges, which is a contradiction to Lemma 16.2(1). □

LEMMA 16.4. *Suppose $w_i = w_j = 4$ and $\hat{e}_i = e_1 \cup ... \cup e_4$ and $\hat{e}_j = e'_1 \cup ... \cup e'_4$ satisfy (i) they have the same label sequence at u_1, and (ii) e_1 is equidistant to e'_1 on Γ_a. Then there exist at least 4 non co-loop edges in the other two non-loop families of Γ_a.*

PROOF. The graph Γ_a is as shown in Figure 16.2 for the case $(i, j) = (1, 2)$. (The proof works in all cases.) Note that e_1 being equidistant to e'_1 implies that e_k is equidistant to e'_k for $k = 1, 2, 3, 4$. We may assume that the label sequence of $\hat{e}_i$ and $\hat{e}_j$ is $1, 2, 3, 4$ at u_1. By Lemma 16.2(1) the four edges $e_1 \cup ... \cup e_4$ form two essential cycles on Γ_b, so any edge on Γ_b with endpoints v_1, v_3 must be parallel to e_1 or e_3. In particular, the edge e'_1 has label pair $(1, 3)$ and hence must be parallel to either e_1 or e_3. Note that two parallel positive edges are equidistant. Since e'_1 is equidistant to e_1 and e_1 is not equidistant to e_3 on Γ_a, it follows that e'_1 is not equidistant to e_3 on Γ_a, therefore by the Equidistance Lemma and the above we see that e'_1 must be parallel to e_1 on Γ_b. Similarly each e'_k is parallel to e_k on Γ_b. Since e'_k and e_k have the same label k at u_1 on Γ_a, they have the same label 1 at v_k in Γ_b, so there must be another edge e''_k between them. By the above e''_k cannot be in $\hat{e}_i \cup \hat{e}_j$, hence they belong to the other two families of non-loop edges in Γ_a, and the result follows. □

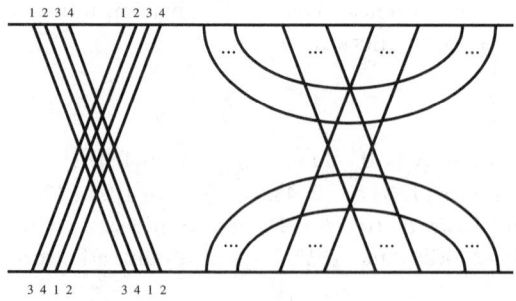

Figure 16.2

LEMMA 16.5. $\lambda = 3$ is impossible.

PROOF. Suppose $\lambda = 3$. Using Lemma 16.2(2)–(3) and the Congruence Lemma 15.1(2) one can show that Γ_a has the following possibilities.
  (1) $\Delta = 5$, $(3; 4, 4, 4, 2)$;
  (2) $\Delta = 4$, $(3; 4, 4, 2, 0)$;
  (3) $\Delta = 4$, $(3; 4, 2, 4, 0)$;
  (4) $\Delta = 4$, $(3; 4, 2, 2, 2)$.

In each case, the family of $\hat{e}_1$ has weight 4. We assume that its label sequence at u_1 is $1, 2, 3, 4$. Then by Lemma 16.2(1) its label sequence at u_2 is $3, 4, 1, 2$, which then completely determines the labels of Γ_a. One can check that in case (1) and (3) the family $\hat{e}_3$ gives 4 parallel co-loops, which is a contradiction to the 3-Cycle Lemma (Lemma 2.14(2)). Case (2) is impossible by Lemma 16.4.

It remains to consider case (4). The graph Γ_a is shown in Figure 16.3. The third edge A of $\hat{e}_1$ and the second edge B of $\hat{e}_3$ in the figure both have label pair $(1, 3)$. As in the proof of Lemma 16.4, this implies that they are parallel on Γ_b. Since $\hat{\Gamma}_a$ has at most 4 negative edges and at most 2 positive edges, by Lemma 2.2(2) Γ_b cannot have more than $2n_a = 4$ parallel edges, so the endpoints of A and B at v_3 are adjacent among the four edge endpoints labeled 1 at v_3. Since $\Delta = 4$, the jumping number is ± 1, so the endpoints of A, B at u_1 in Γ_a are also adjacent among the four edge endpoints labeled 3 at u_1. This is a contradiction because this is not the case in Figure 16.3. □

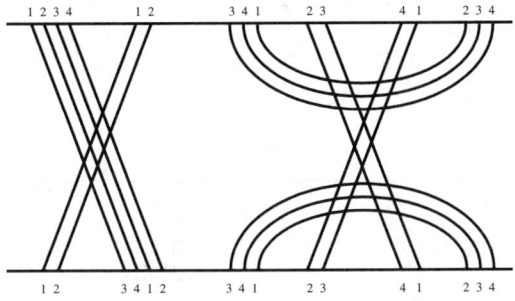

Figure 16.3

LEMMA 16.6. If $\lambda = 4$, then $\Gamma_a = (4; 4, 2, 4, 2)$, and the graphs Γ_a, Γ_b and their edge correspondence are as shown in Figure 16.6.

PROOF. Since Γ_a does not contain an extended Scharlemann cycle, by considering the labels at the endpoints of the four loops at u_1 we see that $w_1 + w_2 \equiv w_3 + w_4 \equiv 2 \pmod 4$. This, together with Lemmas 16.2(2)–(3) and 15.1(2), give the following possibilities for Γ_a.
  (1) $\Delta = 5$, $\Gamma_a = (4; 4, 2, 4, 2)$;
  (2) $\Delta = 5$, $\Gamma_a = (4; 4, 2, 2, 4)$;
  (3) $\Delta = 4$, $\Gamma_a = (4; 4, 2, 2, 0)$.

We shall show that (2) and (3) are impossible, and (1) gives the example in Figure 16.6.

Case (2) can be excluded by Lemma 16.4. The graph Γ_a is shown in Figure 16.4. Note that the corresponding edges of the two non-loop families of weight 4 are equidistant in Γ_a, and they have the same label sequence at u_1. Since the other two non-loop families of Γ_a consist of co-loops, this is a contradiction to Lemma 16.4.

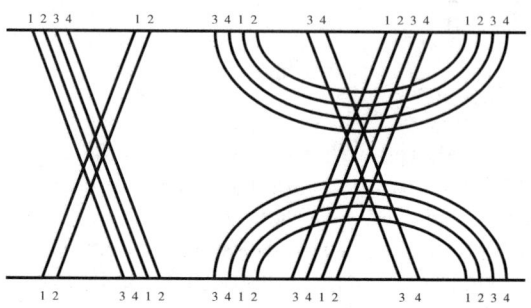

Figure 16.4

The graph for case (3) is shown in Figure 16.5. Note that there is a loop in Γ_b based at each vertex v_i, so two edges connecting v_i to different vertices must be on different sides of the loop. Consider the four edges with label 3 at u_1, indicated by A, B, C, D in Figure 16.5. Note that they appear in this order on ∂u_1. Since $\Delta = 4$, the jumping number is ± 1, so they must also appear in such an order on ∂v_3 in Γ_b.

On the other hand, since A connects v_3 to v_1 while B, D connect v_3 to v_4, A must be on a different side of the loop C at v_3 than B, D. Hence when traveling around ∂v_3 in a certain direction the four edges appear in the order A, C, B, D or A, C, D, B. This is a contradiction. Therefore Case (3) is impossible.

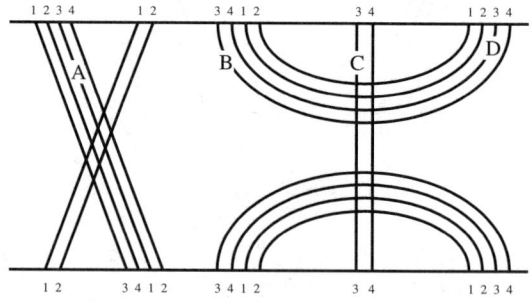

Figure 16.5

In case (1), the graph Γ_a is shown in Figure 16.6(a). By the same argument as above, we see that the edges $B \cup E$ and $A \cup C$ must be on different sides of the loop D in Γ_b. Therefore B, E are adjacent among the 5 edges labeled 1 at v_3. Since they are not adjacent among the 3-edges at u_1, the jumping number must be ± 2. This completely determines the edges around the vertex v_3 up to symmetry, which in turn determine the edges at adjacent vertices v_1, v_3 and then the edges at v_2. The graph Γ_b is shown in Figure 16.6(b). □

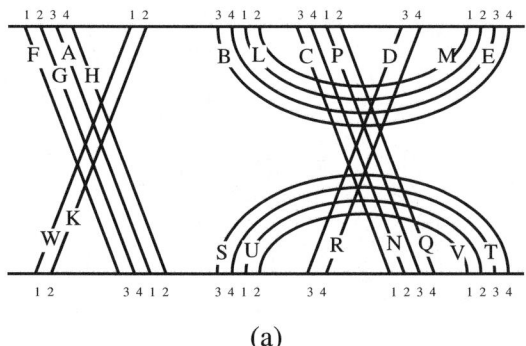

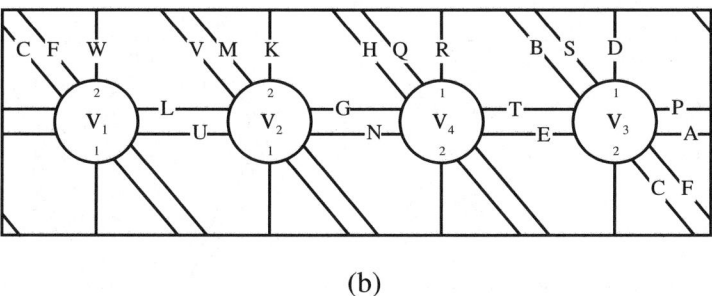

Figure 16.6

LEMMA 16.7. *If $\lambda = 2$, then either $\Delta = 5$ and $\Gamma_a = (2;4,4,4,4)$, or $\Delta = 4$ and $\Gamma_a = (2;4,4,4,0)$. The graphs Γ_a, Γ_b are as shown in Figures 16.8 and 16.9.*

PROOF. Here the possibilities for Γ_a are
(1) $\Delta = 5$, $\Gamma_a = (2;4,4,4,4)$;
(2) $\Delta = 4$, $\Gamma_a = (2;4,4,4,0)$;
(3) $\Delta = 4$, $\Gamma_a = (2;4,4,2,2)$;
(4) $\Delta = 4$, $\Gamma_a = (2;4,2,4,2)$;
(5) $\Delta = 4$, $\Gamma_a = (2;4,2,2,4)$.

The graphs in cases (3) – (5) are shown in Figure 16.7 (a) – (c). In cases (3) and (4) the corresponding edges in the two weight 4 families are equidistant, and the other two non-loop families are co-loops. Therefore these cases are impossible by Lemma 16.4. In case (5) there are loops at v_1 and v_2 in Γ_b, and there is a (34)-Scharlemann bigon in Γ_a which forms another essential cycle C in Γ_b. Consider the two edges of Γ_a with label 3 at u_1 and label 1 at u_2. On Γ_b these edges connect v_3 and v_1, and therefore must lie on the same side of C. Hence they are adjacent among the four edges labeled 1 at v_3 because the other two edges connect v_3 to v_4. Since $\Delta = 4$, the jumping number must be ± 1, so these edges are also adjacent among the four edges with label 3 at u_1, which is a contradiction because on Figure 16.7(c) the two edges with label 3 at u_1 and 1 at u_2 are not adjacent among the four edges labeled 3 at u_1. Therefore (5) is also impossible.

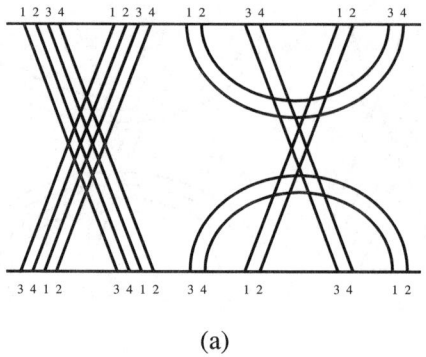

(a)

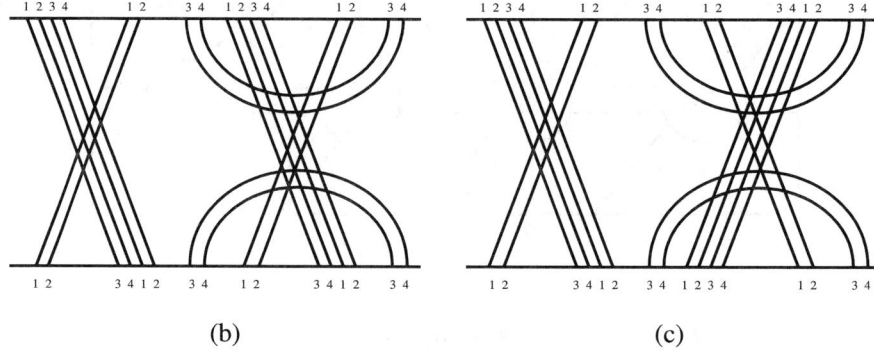

(b) (c)

Figure 16.7

In case (1) the graph Γ_a is shown in Figure 16.8(a). Label the edges as in the figure, and orient non-loop edges of Γ_a from u_1 to u_2. As in the proof of Lemma 16.4, the i-th edge e_i in $\hat{e}_1$ must be parallel to the i-th edge e'_i in $\hat{e}_2$ on Γ_b, and there is an edge of $\hat{e}_3 \cup \hat{e}_4$ between them because e_i, e'_i both have label 1 at v_i. For the same reason the i-th edge of $\hat{e}_3$ is parallel to the i-th edge of $\hat{e}_4$, hence the positive edges of Γ_b form four families of weight 4. The two edges e_i, e'_i are adjacent among the five edges labeled 1 at v_i in Γ_b, hence the jumping number $J = \pm 1$. Reversing the orientation of the vertices of Γ_b if necessary we may assume $J = 1$. We may also assume that the vertices v_1, v_3 are oriented counterclockwise and v_2, v_4 clockwise, otherwise we may look at $\hat{F}_b$ from the other side.

Since Γ_b contains 4 parallel positive edges, by Lemma 13.2(2) Γ_a is kleinian, so the weight of edges of $\hat{\Gamma}_b$ are all even. There are only two (14)-edges K, W in Γ_a, so they must be parallel in Γ_b. They may appear in the order (K, W) or (W, K) on ∂v_1, but there is a homeomorphism of $(\hat{F}_a, \Gamma_a)$ which is label preserving, interchanging u_1, u_2 and mapping K to W, hence up to symmetry we may assume that the order is (K, W). Thus up to symmetry we may assume that K and W appear in Γ_b as shown in Figure 16.8(b). This, together with the orientation of the vertices and the fact that $J = 1$, completely determines the edges around v_1 and v_4, and then the edges around v_2 and v_3. See Figure 16.8(b).

16. THE CASE $n_a = 2$, $n_b = 3$ OR 4, AND Γ_1, Γ_2 NON-POSITIVE   93

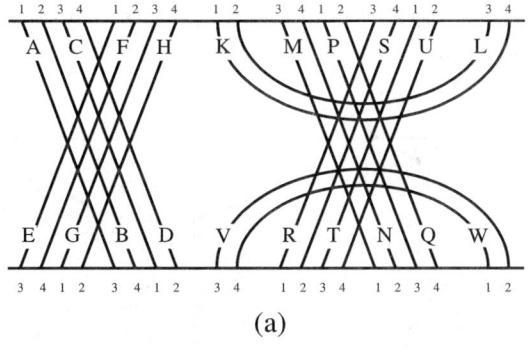

(a)

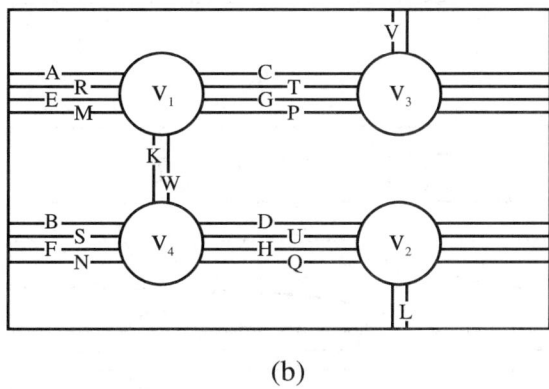

(b)

Figure 16.8

The graph Γ_a in case (2) is shown in Figure 16.9(a). As above, one can show that each edge e_i in $\hat{e}_1$ is parallel to $e_i' \in \hat{e}_2$ and $e_i'' \in \hat{e}_3$, where e_i, e_i' have label i at u_1 and e_i'' has label i at u_2. Orient v_i as above. Up to symmetry we may assume $J = 1$, and A, E on Γ_b are as shown in Figure 16.9(b). This determines P and the position of K at ∂v_1, and hence the labels of the 2-edges at v_1. The 4-labels at u_1 appear in the order K, D, H, N, so on Γ_b they appear in this order around v_4, clockwise, hence D, H must be to the right of v_4 in the figure. This also determines the 2-edges at v_4. In particular, the edges K and S must be non-parallel. The remaining two edges R and L can be determined similarly, using labels at v_2 and v_3. See Figure 16.9(b). □

PROPOSITION 16.8. *Suppose $n_a \leq 2$ and $n_b \geq 3$. Then Γ_a, Γ_b and their edge correspondence are given in Figure 11.9, 11.10, 14.5, 16.6, 16.8 or 16.9.*

PROOF. First assume that Γ_a is positive. Then $n_b \leq 4$ by Lemma 3.2. By Lemma 2.23 n_b must be even, hence our assumption implies that $n_b = 4$. By Proposition 11.9 the graphs are as shown in Figure 11.9 or 11.10.

Now assume Γ_a is non-positive. Then we have $n_a = 2$. The case that Γ_b is positive has been ruled out by Proposition 12.17. Hence Γ_a, Γ_b are both non-positive. By Lemma 16.1 n_b cannot be 3. By Proposition 14.7 if $n_b > 4$ then Γ_a, Γ_b are given in Figure 14.5. Finally if $n_b = 4$. Then Lemma 16.3 and 16.5 says that $\lambda = 4$ or 2, which are covered by Lemmas 16.6 and 16.7, respectively, showing that if $\lambda = 4$ then the graphs are in Figure 16.6, and if $\lambda = 2$ then the graphs are the pair in Figure 16.8 or 16.9. □

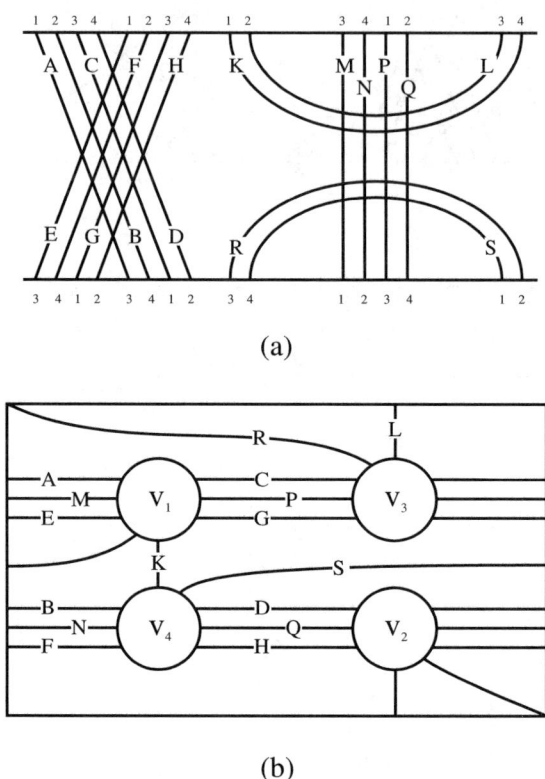

Figure 16.9

17. Equidistance classes

The next few sections deal with the case that $n_i \leq 2$ for $i = 1, 2$. In this section we introduce the concept of equidistance classes. The main properties are given in Lemmas 17.1 and 17.2, which will be used extensively in the next few sections.

Define a relation on the set of edges E_a of Γ_a such that $e_1 \sim e_2$ if and only if (i) they have the same label pair, (ii) they have the same endpoint vertices, and (iii) they are equidistant.

LEMMA 17.1. *This is an equivalence relation.*

PROOF. We need only show that condition (iii) is transitive, i.e. if e_1, e_2, e_3 are edges on a graph Γ such that e_1, e_2 and e_2, e_3 are equidistant pairs, then e_1, e_3 are equidistant.

By definition we have $d_{u_1}(e_1, e_2) = d_{u_2}(e_2, e_1)$, and $d_{u_1}(e_2, e_3) = d_{u_2}(e_3, e_2)$, hence $d_{u_1}(e_1, e_3) = d_{u_1}(e_1, e_2) + d_{u_1}(e_2, e_3) = d_{u_2}(e_2, e_1) + d_{u_2}(e_3, e_2) = d_{u_2}(e_3, e_1)$. This completes the proof. □

We will call this equivalence relation the *ED relation*. An equivalence class is then called an *ED class*, and the number of ED classes is called the *ED number* of

Γ_a, denoted by $\eta_a = \eta(\Gamma_a)$. We can then define $D_a = D(\Gamma_a) = (c_1, ..., c_{\eta_a})$, where c_i are the number of edges of the equivalence classes, ordered lexicographically.

LEMMA 17.2. *Let Γ_a, Γ_b be intersection graphs. Then the edge correspondence between the graphs induces a one to one correspondence between the ED classes of Γ_a and Γ_b; in particular $\eta(\Gamma_a) = \eta(\Gamma_b)$, and $D(\Gamma_a) = D(\Gamma_b)$.*

PROOF. Note that e_1, e_2 satisfy (i) on Γ_a if and only if they satisfy (ii) on Γ_b. The Equidistance Lemma 2.17 now says that a pair of edges are equivalent on Γ_a if and only if they are equivalent on Γ_b. □

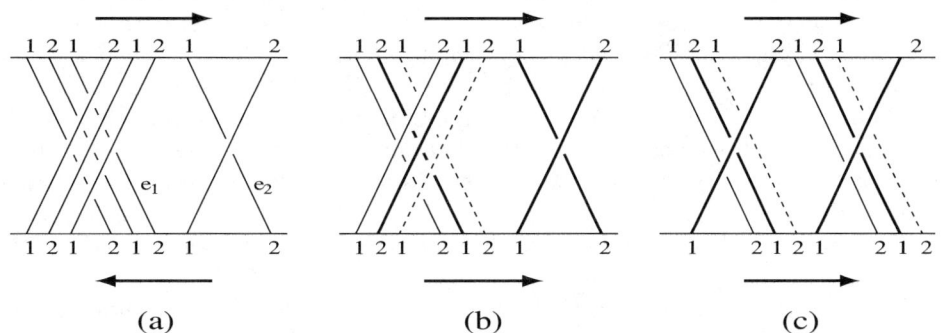

Figure 17.1

Example 17.3 (1) Consider a graph $\Gamma_a = +(3, 3, 1, 1)$ and assume $n_b = 2$, see Figure 17.1(a). In general if $n_b = 2$ then all parallel positive edges are in the same ED class because they have the same label pairs and they are equidistant. One can check that non-parallel edges are not equidistant. (For example, let u_1 be the top vertex, u_2 the bottom vertex, and let e_1, e_2 be as shown in the figure; then $d_{u_1}(e_1, e_2) = 4 \neq 2 = d_{u_2}(e_2, e_1)$.) Hence $D(\Gamma_a) = (3, 3, 1, 1)$. Compare this with $\hat{\Gamma}_a = +(3, 1, 3, 1)$, in which case the two families of 3 edges are equidistant, and the other two families of weight 1 are equidistant, hence $D(+(3, 1, 3, 1)) = (6, 2)$.

(2) Consider $\Gamma_a = -(3, 3, 1, 1)$ or $-(3, 1, 3, 1)$, and suppose that the edges of Γ_a are not co-loops (hence conditions (i) and (ii) in the definition of ED equivalence are satisfied), see Figure 17.1(b) and (c). Equidistant edges are indicated in the figure by different kind of lines. We can see that $D(-(3, 3, 1, 1)) = D(-(3, 1, 3, 1)) = (4, 2, 2)$.

(3) When $\Gamma_a = -(4, 2, 2, 0)$, each of the middle edges of the family of 4 is equidistant to one edge in each of the two weight 2 families, and the other two edges of the weight 4 families are not equidistant to any other edges. Hence $D(-(4, 2, 2, 0)) = (3, 3, 1, 1)$

(4) Suppose $\Gamma_a = +(4, 2, 2, 0)$ and all edges have label pair (12). Then one can check that each family of parallel edges forms an ED class, hence $D(\Gamma_a) = (4, 2, 2)$.

(5) Suppose $\Gamma_a = +(2, 2, 2, 2)$ and all edges have label pair (12). Then one can show that the first family is equidistant to the third family, but not to the adjacent families. Hence $D(\Gamma_a) = (4, 4)$.

(6) Similarly if $\Gamma_a = +(4, 4, 0, 0)$ and all edges have the same label pair then $D(\Gamma_a) = (8)$.

18. The case $n_b = 1$ and $n_a = 2$

LEMMA 18.1. *Suppose $n_a = 2$ and $n_b = 1$. Then one of the following holds.*
(1) $\Gamma_a = -(1,1,1,1)$ and $\Gamma_b = (4,0,0)$.
(2) $\Gamma_a = -(2,2,0,0)$ and $\Gamma_b = (2,2,0)$.
(3) $\Gamma_a = -(2,1,1,1)$ and $\Gamma_b = (3,1,1)$. The graphs Γ_a, Γ_b and their edge correspondence are given in Figure 18.2.

PROOF. In this case $\hat{\Gamma}_b$ has a single vertex, and $\hat{\Gamma}_a$ has two vertices of opposite orientation and has no loops. Hence we have $\Gamma_a = (a_1, ..., a_4)$, and $\Gamma_b = (b_1, b_2, b_3)$. We have $b_1 + b_2 + b_3 = \Delta$. If b_i, b_j are non-zero and $b_i + b_j$ is odd then one can check that one of the $\hat{e}_i, \hat{e}_j$ is a family of co-loops, which is a contradiction to the parity rule. Hence $b_i \equiv b_j \mod 2$ for all b_i, b_j non-zero. Thus if $\Delta = 5$ then up to symmetry we have $\Gamma_b = (3,1,1)$ or $(5,0,0)$, and if $\Delta = 4$ then $\Gamma_b = (4,0,0)$, $(2,2,0)$ or $(3,1,0)$. The case $(5,0,0)$ is impossible because two of the edges would be parallel on both graphs, contradicting Lemma 2.2(2). The case $(3,1,0)$ can be ruled out by the parity rule, since in this case a loop would have the same label at its two endpoints.

If $\Gamma_b = (4,0,0)$ then the four parallel edges are mutually non-parallel on Γ_a, hence $\Gamma_a = -(1,1,1,1)$.

If $\Gamma_b = (2,2,0)$, one can check that edges in different families are not equidistant, hence $D(\Gamma_b) = (2,2)$. Since each pair of parallel edges contributes one edge to each of two families in Γ_a, we have $\Gamma_a = -(2,2,0,0), -(2,1,1,0)$ or $-(1,1,1,1)$. When $\Gamma_a = -(2,1,1,0)$ the two single edges are equidistant, while each of the two parallel edges form an ED class, so $D(\Gamma_a) = (2,1,1) \neq D(\Gamma_b)$. Also, when $\Gamma_a = -(1,1,1,1)$ we have $D(\Gamma_a) = (4)$. Therefore in this case we have $\Gamma_a = -(2,2,0,0)$.

No suppose $\Gamma_b = (3,1,1)$. In this case the three parallel edges are equidistant, and each of the other two edges is not equidistant to any other edges. Hence $D(\Gamma_b) = (3,1,1)$. Since the three parallel edges in Γ_b are mutually non-parallel on Γ_a, $\hat{\Gamma}_a$ has at least three edges. One can show that $D(-(2,2,1,0)) = (2,2,1) \neq D(\Gamma_b)$, hence $\hat{\Gamma}_a \neq -(2,2,1,0)$. Therefore $\Gamma_a = -(3,1,1,0)$ or $-(2,1,1,1)$.

In the case that $\Gamma_a = -(3,1,1,0)$ and $\Gamma_b = (3,1,1)$, the graphs are as shown in Figure 18.1. The three parallel edges B, C, E are equidistant, hence they represent the two weight 1 edges $\hat{e}_2, \hat{e}_3$ and the middle edge of the weight 3 edge $\hat{e}_1$, so the other two edges A, D must be as shown in Figure 18.1(a) up to symmetry. Since they are non-adjacent at u_1 and their label 1 endpoints are non-adjacent among the five label 1 edge endpoints at v_1 in Γ_b, the jumping number must be ± 1. This determines the edge correspondence between Γ_a and Γ_b, as shown in Figure 18.1.

The torus $\hat{F}_a$ cuts $M(r_a)$ into two components. Let W be the one containing the bigon α on F_b bounded by $B \cup E$ and the 3-gon β bounded by $A \cup C \cup D$. It can be constructed by attaching a 1-handle representing part of the Dehn filling solid torus, then two 2-handles represented by α, β, then a 3-cell. The fundamental group of W is generated by the horizontal circle x and the vertical circle y shown in the figure, and the 1-handle z from u_2 to u_1. On the boundary of $\alpha, \beta, A, B, C, D, E$ represent $1, x, xy, 1, 1$, respectively, and each corner represents z, hence α, β give the relations $zzx = 1$ and $zxyzz = 1$, respectively. Solving these in x and y shows that $\pi_1(W) = \mathbb{Z}$, generated by z. It follows that $\hat{F}_a$ is not π_1-injective in W, and hence is compressible. This is a contradiction.

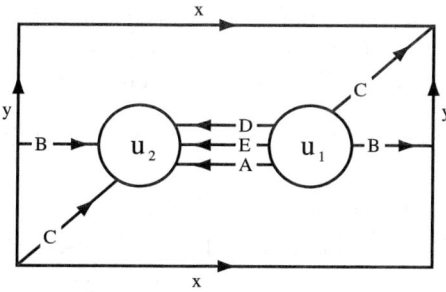 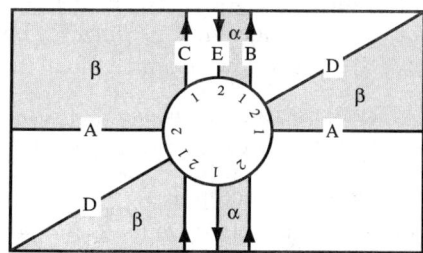

Figure 18.1

We now have $\Gamma_a = -(2,1,1,1)$ and $\Gamma_b = (3,1,1)$. The three parallel edges B, C, E are equidistant, hence on Γ_a they are the single edges because they are equidistant to each other but not to the edges in the weight 2 family. Since the edge endpoints of these are consecutive on ∂v_1 while the 1-label endpoint of E at v_1 is not adjacent to that of either B or C, the jumping number must be ± 2. This determines the correspondence of the edges up to symmetry, see Figure 18.2. □

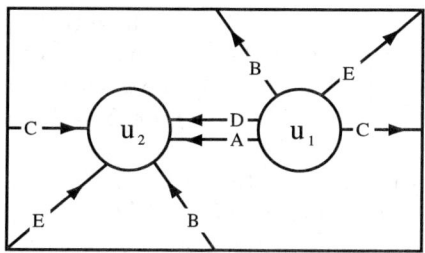 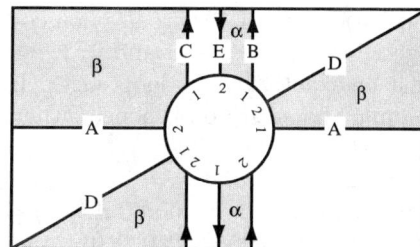

Figure 18.2

19. The case $n_1 = n_2 = 2$ and Γ_b positive

In this section we assume that $n_1 = n_2 = 2$ and Γ_b is positive. Then no edge of $\hat{\Gamma}_a$ is a loop, hence $\Gamma_a = -(a_1, ...a_4)$, and $\Gamma_b = +(\rho; b_1, ..., b_4)$.

When $\rho \neq 0$ we may rearrange the a_i to write $\Gamma_a = -(r_1,...,r_p \,|\, s_1,...,s_q)$, where r_i are the weights of the co-loop edges, and s_j are the weights of the non co-loop edges.

LEMMA 19.1. *Suppose $n_1 = n_2 = 2$ and Γ_b is positive.*

(1) All non-zero b_i are of the same parity. All non-zero r_j are of the same parity, all non-zero s_k are of the same parity, and the non-zero r_j and s_k are of opposite parity.

(2) $r_i \leq 2$, $s_j \leq 4$, and $\rho + b_k \leq 4$.

(3) $2\rho + \sum b_i = 2\Delta$, and $\sum r_i + \sum s_j = 2\Delta$.

PROOF. (1) The follows from the Congruence Lemma 15.1.

(2) Since $\hat{\Gamma}_b$ has at most two loops, each co-loop family of Γ_a contains at most two edges, hence $r_i \leq 2$. Similarly, since $\hat{\Gamma}_b$ has at most four non-loop edges,

$s_j \leq 4$. On Γ_b, ρ is the number of edges in a loop family, which is no more than p, the number of co-loop edges in $\hat\Gamma_a$. Similarly, b_k is no more than q, the number of non co-loop edges in $\hat\Gamma_a$. Since $\hat\Gamma_a$ has at most 4 edges, we have $\rho + b_k \leq 4$.

(3) This follows from the fact that each vertex of Γ_a or Γ_b has valence 2Δ. □

LEMMA 19.2. *Suppose $n_1 = n_2 = 2$ and Γ_b is positive. If $\rho = 4$ then $\Gamma_a = -(2,2,2,2)$, and $\Gamma_b = +(4; 0, 0, 0, 0)$.*

PROOF. Since $\rho + b_j \leq 4$ (Lemma 19.1(2)), we have $b_i = 0$ for all i, hence from Lemma 19.1(3) we have $\Delta = 4$. Thus $\hat\Gamma_b$ is a union of two disjoint loops, each representing a family of four edges. Since each family of four parallel edges in $\hat\Gamma_b$ contributes one edge to each family in Γ_a, we have $\Gamma_a = -(2, 2, 2, 2)$. □

LEMMA 19.3. *Suppose $n_1 = n_2 = 2$ and Γ_b is positive. Then $\rho \neq 3$.*

PROOF. Suppose $\rho = 3$. The three loops in a family represent different classes on $\hat\Gamma_a$, so $\hat\Gamma_a$ has at least three co-loop edges. Since Γ_b has some non-loop edges, $\hat\Gamma_a$ has at least one non co-loop edge. It follows that $\hat\Gamma_a$ has exactly three co-loop edges, so $\Gamma_a = -(2, 2, 2 \,|\, s_1)$. Since $\sum r_i + \sum s_j = 2\Delta$ is even, s_1 is even, which contradicts Lemma 19.1(1). □

LEMMA 19.4. *Suppose $n_1 = n_2 = 2$ and Γ_b is positive. Then $\rho \neq 2$.*

PROOF. On $\hat\Gamma_a$ there are non co-loop edges, so there are at most three co-loop edges, but since $r_i \leq 2$ and $\sum r_i = 4$ and the r_i's are of the same parity, there must be exactly two co-loop edges. Hence $\Gamma_a = -(2, 2 \,|\, s_1, s_2)$. By the Congruence Lemma, the s_i are odd, hence either $\Delta = 5$ and $\Gamma_a = -(2, 2 \,|\, 3, 3)$, or $\Delta = 4$ and $\Gamma_a = -(2, 2 \,|\, 3, 1)$.

If $\Delta = 4$ and $\Gamma_a = -(2, 2 \,|\, 3, 1)$, then from Lemma 19.1 we have $b_i \leq 2$, $\sum b_i = 4$, and $b_i \equiv b_j \bmod 2$ if $b_i, b_j \neq 0$. These conditions give $\Gamma_b = +(2; 1, 1, 1, 1)$, $+(2; 2, 2, 0, 0)$ or $+(2; 2, 0, 2, 0)$. One can check that in the first two cases the four non-loop edges of Γ_b form two equidistance classes of 2 edges each, so $D_b = (2, 2, 2, 2)$, and in the third case the four non-loop edges are all equidistant to each other, so $D_b = (4, 2, 2)$. On the other hand, the three parallel edges of Γ_a belong to distinct classes, and there are at least two co-loop classes, hence $\eta(\Gamma_a) \geq 5$. This is a contradiction to Lemma 17.2.

If $\Delta = 5$ and $\Gamma_a = -(2, 2 \,|\, 3, 3)$, then from Lemma 19.1 we have $b_i \leq 4 - \rho = 2$, $\sum b_i = 6$, and $b_i \equiv b_j \bmod 2$ if $b_i, b_j \neq 0$, so we must have $\Gamma_b = +(2; 2, 2, 2, 0)$. Depending on the weight sequence of the edges of $\hat\Gamma_a$, we have $\Gamma_a = -(3, 2, 3, 2)$ or $-(3, 3, 2, 2)$. If $\Gamma_a = -(3, 2, 3, 2)$ then from the labeling one can see that the two edges with both endpoints labeled 1 are not equidistant. Since these are parallel loops at v_1 of Γ_b, they are equidistant on Γ_b, which is a contradiction to the Equidistance Lemma 2.17. Therefore $\Gamma_a \neq (-3, 2, 3, 2)$.

Now suppose $\Gamma_a = -(3, 3, 2, 2)$, and $\Gamma_b = +(2; 2, 2, 2, 0)$. Then the graphs are as shown in Figure 19.1. Consider the edges A, B, C, D, E with label 1 at u_1 of Γ_a. These correspond to the 5 edges with label 1 at v_1 of Γ_b. Note that on Γ_a D, E are co-loop edges, hence on Γ_b they are the two loops at v_1. Since their endpoints with label 1 are not adjacent among the 1-edges at v_1 in Γ_b, the jumping number must be ± 2, so among these edges in Γ_b, the edge B is the one in Γ_b which is adjacent to both D and E at v_1, as shown in the Figure. Now consider the five edges labeled 2 at u_2. Note that they appear in the order $CABFG$. Using the same argument as above we see that the edge A is the one adjacent to both F and G, so we would have $A = B$, which is a contradiction. □

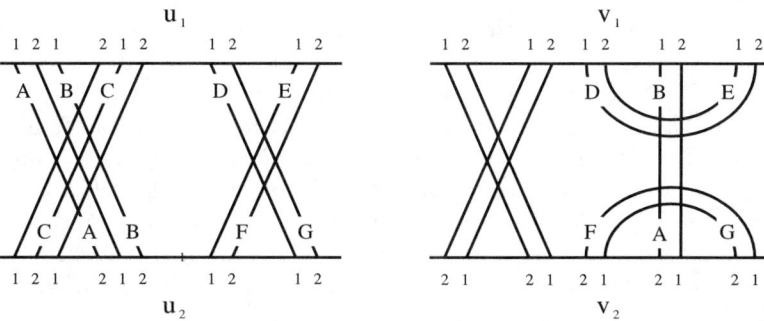

Figure 19.1

LEMMA 19.5. *Suppose $n_1 = n_2 = 2$ and Γ_b is positive. Then $\rho \neq 1$.*

PROOF. Suppose $\rho = 1$. Then Γ_a has two co-loop edges, hence $\Gamma_a = -(1,1 \,|\, s_1, s_2)$ or $-(2 \,|\, s_1, s_2, s_3)$.

If $\Delta = 5$ then the second case does not happen since by the Congruence Lemma s_i would be 0 or odd and $\sum s_i = 8$, which would give $s_i > 4$ for some i, contradicting Lemma 19.1(2). Therefore $\Gamma_a = -(1,1 \,|\, 4, 4)$. Since each weight 4 family contributes one edge to each non-loop family of Γ_b, we have $\Gamma_b = +(1; 2, 2, 2, 2)$. Now the graph Γ_b contains both black and white bigons, whose edges all belong to the two weight 4 families on Γ_a. This is a contradiction to Lemma 15.2.

If $\Delta = 4$ then $\Gamma_a = -(1, 1 \,|\, 4, 2)$ or $-(2 \,|\, 3, 3, 0)$. The first case cannot happen because the $\hat{e}_i$ of weight 4 contributes one edge to each family in $\hat{\Gamma}_b$, while the edge of weight 2 contributes one edge to each of two families, so $\Gamma_b = +(1; 2, 2, 1, 1)$ or $+(1; 2, 1, 2, 1)$, which contradicts the Congruence Lemma. In the second case for the same reason above we must have $\Gamma_b = +(1; 2, 2, 2, 0)$. Again there are black and white bigons, which contradicts Lemma 15.2 because on Γ_a the edges of these bigons all belong to the two weight 3 families. □

LEMMA 19.6. *Suppose $n_1 = n_2 = 2$ and Γ_b is positive. If $\rho = 0$ then $\Delta = 5$.*

PROOF. In this case there is no loop on either graph, hence by Lemma 19.1(1) all non-zero a_i have the same parity, and all non-zero b_j have the same parity. Any two edges connect the same pair of vertices and have the same pair of labels on their two endpoints, hence by definition they are ED equivalent if and only if they are equidistant.

Assume $\Delta = 4$. By the Congruence Lemma each of Γ_a and Γ_b is of type $(4, 4, 0, 0)$, $(2, 2, 2, 2)$, $(4, 2, 2, 0)$, $(3, 1, 3, 1)$, or $(3, 3, 1, 1)$. Let $e_1 \cup e_2$ be a bigon on Γ_b. Then e_1 and e_2 are equidistant on Γ_b, so by Lemma 2.17 they form an equidistant pair on Γ_a. Note that since e_1 and e_2 are not loops on Γ_b, on Γ_a they have different labels on u_1. On the other hand, one can check that if $\Gamma_a = -(4, 4, 0, 0)$ or $-(2, 2, 2, 2)$ then an equidistant pair e_1, e_2 on Γ_a must have the same label on u_1, which is a contradiction. Therefore $\Gamma_a = -(4, 2, 2, 0)$, $-(3, 3, 1, 1)$ or

$-(3, 1, 3, 1)$. (Note that the above argument does not apply to Γ_b since a pair of parallel edges on Γ_a is not an equidistant pair.) We will rule these out one by one.

CLAIM 1. *The case $\Gamma_a = -(4, 2, 2, 0)$ is impossible.*

If $\Gamma_a = -(4, 2, 2, 0)$ then $b_i \neq 0$ for all i, hence $\Gamma_b = +(2, 2, 2, 2)$ or $+(3, 1, 3, 1)$, or $+(3, 3, 1, 1)$. In the first case all black (say) faces of Γ_b are bigons, so by Lemma 13.2(1) the 8 edges form either 2 or at least 4 families of edges in Γ_a, contradicting the fact that $\Gamma_a = -(4, 2, 2, 0)$. Also, by Example 17.3 we have $D(-(4, 2, 2, 0)) = (3, 3, 1, 1)$, and $D(+(3, 1, 3, 1)) = (6, 2)$, hence $\Gamma_b \neq +(3, 1, 3, 1)$. It follows that $\Gamma_b = +(3, 3, 1, 1)$. The graphs are shown in Figure 19.2. Each of A, D on Γ_a forms an equidistance class, hence they are the single edges on Γ_b. Up to symmetry we may assume that A, D are as shown in Figure 19.2(b). This and the jumping number J determine the edge correspondence of the graphs. The case that $J = 1$ is shown in the figure. When $J = -1$ the edges G, E would be equidistant on Γ_a but not on Γ_b, which is impossible.

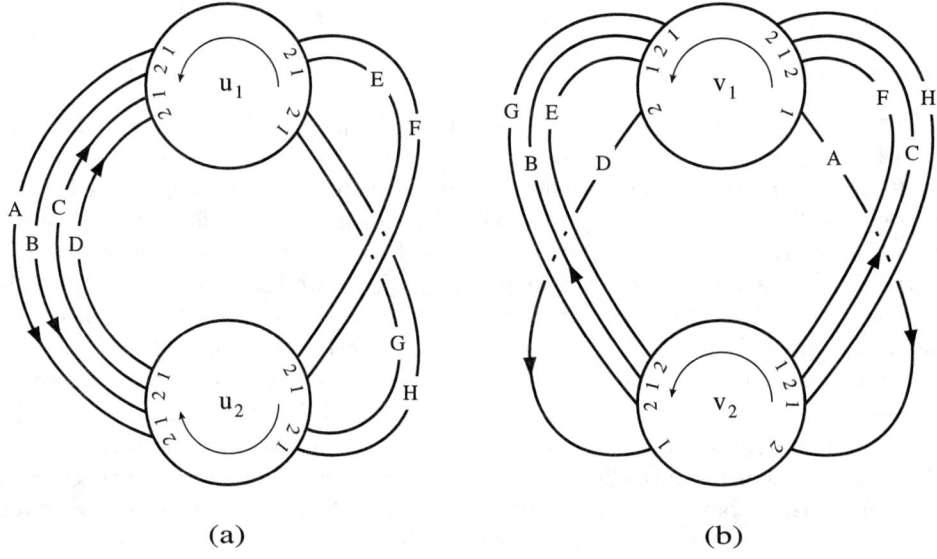

Figure 19.2

Let P_1, P_2 be the bigon disks on F_a bounded by $A \cup B$ and $C \cup D$, respectively. Then the union of P_1, P_2 and two disks on T_0 form an annulus Q. More explicitly, let a_1 (resp. b_1) be the arc on ∂u_1 (resp. ∂u_2) from the endpoint of A to that of B, a_2 (resp. b_2) the arc on ∂u_1 (resp. ∂u_2) from the endpoint of C to that of D, a_3 (resp. b_3) on ∂v_1 (resp. ∂v_2) from A to C, and a_4 (resp. b_4) on ∂v_2 (resp. ∂v_1) from B to D. Then $a_1 \cup ... \cup a_4$ (resp. $b_1 \cup ... \cup b_4$) bounds a disk P_3 (resp. P_4) on the boundary torus T_0. Now $Q = P_1 \cup ... \cup P_4$ is an annulus in M. Note that ∂Q consists of two simple closed curves $\partial_1 = A \cup C \cup a_3 \cup b_3$ and $\partial_2 = B \cup D \cup a_4 \cup b_4$.

Orient A, B to point from u_1 to u_2 on Γ_a. This determines the orientation of ∂_i. Note that they are parallel on the annulus Q. On Γ_b the orientations of A, B are from label 1 to label 2, as shown in the figure. This determines the orientations of C, D. It is important to see that ∂_1, ∂_2 are parallel as oriented curves on $\hat{F}_b$. Let

Q' be an annulus on $\hat{F}_b$ with $\partial Q' = \partial_1 \cup \partial_2$. Then $Q \cup Q'$ is a non-separating torus (not a Klein bottle!) in $M(r_b)$ intersecting the Dehn filling solid torus at a single meridian disk, which contradicts the choice of $\hat{F}_b$. Therefore this case is impossible.

CLAIM 2. *The case $\Gamma_a = -(3,1,3,1)$ is impossible. If $\Gamma_a = -(3,3,1,1)$ then $\Gamma_b = +(4,2,2,0)$.*

Now suppose $\Gamma_a = -(3,3,1,1)$ or $-(3,1,3,1)$ and $\Gamma_b = +(4,4,0,0), +(4,2,2,0),$ $+(2,2,2,2), +(3,3,1,1)$ or $+(3,1,3,1)$. By Example 17.3 we have $D(-(3,3,1,1)) = -(3,1,3,1) = (4,2,2)$. On the other hand, by Example 17.3 we also have $D(+(4,4,0,0)) = (8), D(+(4,2,2,0)) = (4,2,2), D(+(2,2,2,2)) = (4,4), D(+(3,3,1,1)) = (3,3,1,1),$ and $D(+(3,1,3,1)) = (6,2)$. Therefore by Lemma 17.2 in this case we must have $\Gamma_b = +(4,2,2,0)$. If $\Gamma_a = -(3,1,3,1)$ then the four edges in the same ED class all have label 2 (say) at u_1, which means that on Γ_b they all have label 1 at v_2, so they cannot be the four parallel edges in $+(4,2,2,0)$. Therefore $\Gamma_a \neq -(3,1,3,1)$.

CLAIM 3. *The case $\Gamma_a = -(3,3,1,1)$ is impossible.*

By Claim 2 we have $\Gamma_b = +(4,2,2,0)$. The graphs are as shown in Figure 19.3. While the graphs are similar to those in Figure 19.2, the argument is necessarily different because the orientations of the vertices of Γ_b here are parallel while those of Γ_a in Figure 19.2 are antiparallel. One can check that up to symmetry the edge correspondence must be as shown in the figure.

We would like to apply Lemma 2.15 to get a contradiction. To do that, let Q be the face of Γ_a bounded by $A \cup B \cup E \cup H$. The edge B is parallel to C on Γ_b, and C is a non-border edge on Γ_a, hence one of the bigons $C \cup H$ or $C \cup F$ is a coupling face Q' of Q along the edge B. By Lemma 2.15 there is a rel ∂ isotopy of F_a such that the new intersection graph Γ'_a is obtained from Γ_a by deleting A and E and adding two edges parallel to B and H, respectively. It follows that $\Gamma'_a = -(4,2,2,0)$. This is impossible by Claim 1. Therefore the case $\Gamma_a = -(3,3,1,1)$ is also impossible. □

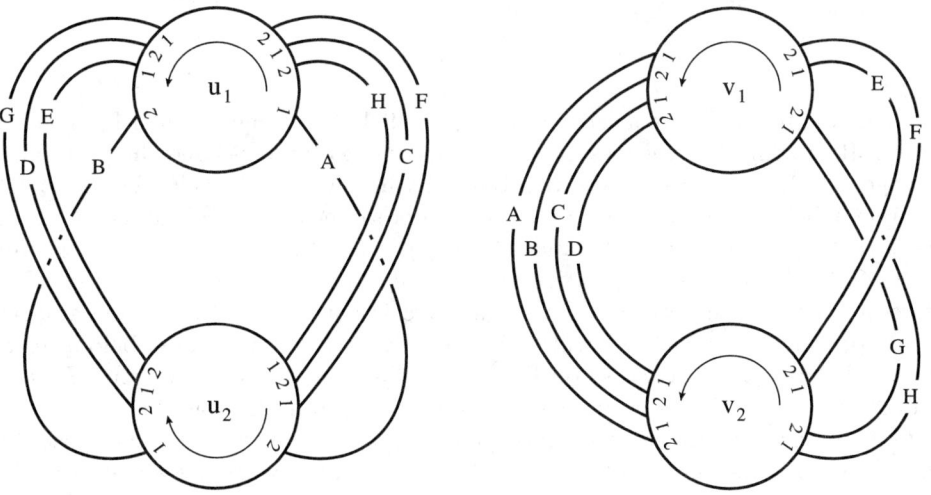

Figure 19.3

LEMMA 19.7. *Suppose* $n_1 = n_2 = 2$ *and* Γ_b *is positive. If* $\rho = 0$ *then* $\Gamma_a = -(3,3,3,1)$ *and* $\Gamma_b = +(3,3,3,1)$. *The graphs* Γ_a, Γ_b *and their edge correspondence are shown in Figure 19.4.*

PROOF. By Lemma 15.1 all non-zero a_i have the same parity, and all non-zero b_j have the same parity. By Lemma 19.6 we have $\Delta = 5$, so each of Γ_a and Γ_b is of type $(4,4,2,0)$, $(4,2,2,2)$ or $(3,3,3,1)$. If some $b_i = 4$ then by Lemma 13.2(2) Γ_a is kleinian, but since each of the above type has an edge whose weight is non-zero and different from the others, it must be mapped to itself by the involution in Lemma 6.2(4), which is a contradiction because it is supposed to be a free involution on $\hat{F}_a$. It follows that $\Gamma_b = +(3,3,3,1)$. Direct calculation gives $D(+(3,3,3,1)) = (4,3,3)$, $D(-(4,4,2,0)) = (3,3,2,2)$, $D(-(4,2,2,2)) = (4,4,1,1)$, and $D(-(3,3,3,1)) = (4,3,3)$. Hence by Lemma 17.2 we have $\Gamma_a = -(3,3,3,1)$.

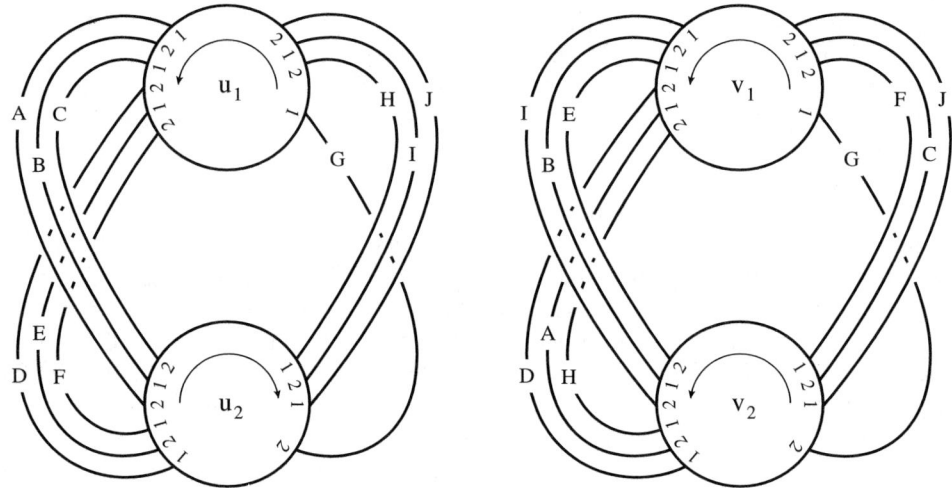

Figure 19.4

The graphs Γ_a, Γ_b are as shown in Figure 19.4. Label the edges of Γ_a as in the figure. Relabeling the vertices of Γ_b if necessary, we may assume that the labels of edges of Γ_a at u_1 are as shown. Since Γ_b has no loops, each edge of Γ_a has different labels on its two endpoints, which determines the labels at u_2. One can check that Γ_a has three equidistance classes $c_1 = \{B, E, G, I\}$, $c_2 = \{A, D, H\}$ and $c_3 = \{C, F, J\}$. Since Γ_b is positive, each family belongs to an equidistance class; moreover, one can check that the single edge is equidistant to the non-adjacent family of weight 3, which we will denote $\hat{e}_1$. Therefore these must belong to c_1. On Γ_a, B has label 1 at u_2, so on Γ_b B has label 2 at v_1. It follows that B is the middle edge in $\hat{e}_1$. This determines the labels on v_1 and v_2. Now the endpoints of $E, G, I \in c_1$ are adjacent on ∂u_1 among edge endpoints labeled 1 (in the sense that one of them is adjacent to both of the other two), but they are not adjacent on ∂v_1 because the single edge is not adjacent to those in $\hat{e}_1$ among edges with label 1 at v_1 in Γ_b. Therefore the jumping number J cannot be ± 1, so $J = \pm 2$. Reversing the orientations of v_1, v_2 if necessary we may assume that $J = 2$. Thus the edges E, G, I in Γ_b must be as shown. The other edges are now determined by this information.

For example, the edges with label 2 at u_1 appear in the order B, D, F, H, J, so on Γ_b the edges with label 1 at v_2 appear in the order B, H, D, J, F. □

20. The case $n_1 = n_2 = 2$ and both Γ_1, Γ_2 non-positive

In this section we assume that $n_1 = n_2 = 2$ and both Γ_1, Γ_2 are non-positive. Let $\Gamma_a = (\rho_a; a_1, ..., a_4)$, and $\Gamma_b = (\rho_b; b_1, ..., b_4)$. Without loss of generality we may assume that $\rho_b \geq \rho_a$.

LEMMA 20.1. *Suppose $n_1 = n_2 = 2$, and Γ_1, Γ_2 are non-positive.*
(1) $\Delta/2 \leq \rho_b \leq 4$.
(2) $2\rho_a + \sum a_i = 2\Delta$, and $2\rho_b + \sum b_i = 2\Delta$.
(3) $a_i, b_i \leq 2$.

PROOF. (1) Since a loop in Γ_a corresponds to a non-loop in Γ_b and vice versa, we have $\rho_a + \rho_b = \Delta$. We have assumed $\rho_b \geq \rho_a$, so $\rho_b \geq \Delta/2$. Since no two edges are parallel on both graphs and $\hat{\Gamma}_a$ has at most four non-loop edges, we also have $\rho_b \leq 4$.

(2) This follows from the fact that the valence of a vertex in Γ_a or Γ_b is 2Δ.

(3) Since Γ_a and Γ_b are non-positive, a non-loop edge in Γ_a is a loop in Γ_b, hence there are at most two edges in each non-loop family of Γ_a, i.e. $a_i \leq 2$. Similarly for b_i. □

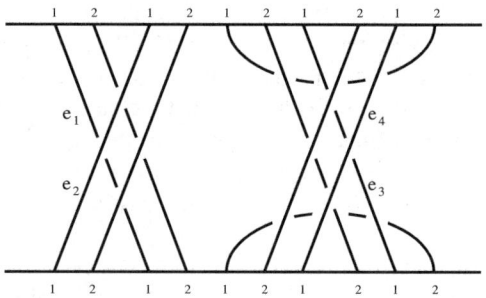

Figure 20.1

LEMMA 20.2. *Suppose $n_1 = n_2 = 2$, $\Delta = 5$, and Γ_1, Γ_2 are non-positive. Then $\Gamma_a = -(2; 2, 2, 2, 0)$ and $\Gamma_b = -(3; 1, 1, 1, 1)$ or $-(3; 2, 2, 0, 0)$.*

PROOF. By Lemma 20.1(1) we have $\rho_b = 3$ or 4. If $\rho_b = 4$ then each loop family contributes one edge to each non-loop family of Γ_a, hence $\Gamma_a = -(1; 2, 2, 2, 2)$. The four loops e_1, e_2, e_3, e_4 at v_1 are equidistant to each other; on the other hand, from Figure 20.1 one can see that e_i is equidistant to e_j on Γ_a if and only if e_i and e_j are on the same side of the loop at u_1. This is a contradiction. Therefore this case cannot happen.

Now assume $\rho_b = 3$. Then by the Congruence Lemma we have $\Gamma_b = -(3; 2, 0, 2, 0)$, $-(3; 2, 2, 0, 0)$ or $-(3; 1, 1, 1, 1)$. Since $\rho_a = \Delta - \rho_b = 2$, we have $\sum a_i = 10 - 2\rho_a = 6$. By Lemma 20.1 we have $a_i \leq 2$, therefore by the Congruence Lemma we must have $\Gamma_a = -(2; 2, 2, 2, 0)$. The first case for Γ_b above cannot happen because the two non-loop 1-edges are not equidistant in Γ_b while as parallel loops on Γ_a they are equidistant on Γ_a. Therefore $\Gamma_b = -(3; 2, 2, 0, 0)$ or $-(3; 1, 1, 1, 1)$. □

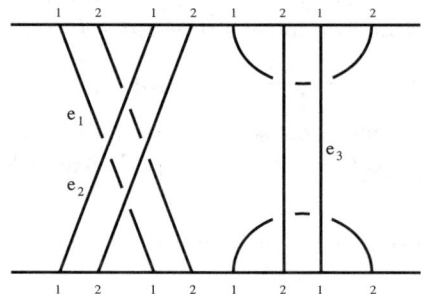

Figure 20.2

LEMMA 20.3. *Suppose $n_1 = n_2 = 2$, $\Delta = 4$, and Γ_1, Γ_2 are non-positive. Then one of the following holds.*
  (1) $\Gamma_a = -(2,2,2,2)$ and $\Gamma_b = -(4;0,0,0,0)$.
  (2) Both Γ_1 and Γ_2 are of type $-(2;1,1,1,1)$.
  (3) Both Γ_1 and Γ_2 are of type $-(2;2,0,2,0)$.
  (4) Both Γ_1 and Γ_2 are of type $-(2;2,2,0,0)$.

PROOF. If $\rho_b = 4$ then $\Gamma_b = -(4;0,0,0,0)$ and each loop family contributes one edge to each family of Γ_a, hence $\Gamma_a = -(2,2,2,2)$.

If $\rho_b = 3$, then by the Congruence Lemma 15.1(2) we have $\Gamma_a = -(1;2,2,2,0)$. Let e_1, e_2, e_3 be the three loops at v_1. As parallel positive edges, they are equidistant on Γ_b. On Γ_a they are as shown in Figure 20.2. One can check that e_1 is equidistant to e_2 but not e_3, which is a contradiction to Lemma 2.17. Therefore $\rho_b \neq 3$.

When $\rho_b = \rho_a = 2$, by the Congruence Lemma each of Γ_a and Γ_b is of type $-(2;1,1,1,1)$ or $-(2;2,2,0,0)$ or $-(2;2,0,2,0)$. We are done if both Γ_a, Γ_b are of the same type.

If $\Gamma_a = -(2;2,2,0,0)$ then the two non-loop edges with label 1 at both endpoints are adjacent among the four edges labeled 1 at u_1, hence on Γ_b the two loops at u_1 are adjacent among the four edges with label 1 at v_1, which implies that Γ_b cannot be $-(2;2,0,2,0)$ or $-(2;1,1,1,1)$.

It remains to rule out the possibility that $\Gamma_a = -(2;1,1,1,1)$ and $\Gamma_b = -(2;2,0,2,0)$. In this case the graphs are as shown in Figure 20.3.

Label the edges of Γ_a as in the figure. We want to show that this determines the labels of the edges of Γ_b up to symmetry. Since $\Delta = 4$, by changing the orientation of $\hat{F}_b$ if necessary we may assume that the jumping number is 1. The 1-edges at u_1 are in the order A, B, C, D, so these labels appear in this order at v_1 on Γ_b. The order of the 1-edges at u_2 is A, X, C, Y, so the 2-edges at v_1 are also in this order, which determines the edges X, Y on Γ_b. Finally, the order of the 2-edges at u_1 determines the edges E, F in Γ_b. Hence the labels of the graphs are as shown in Figure 20.3.

One way to see that these graphs are not realizable is to consider the annulus A from ∂v_1 to ∂v_2 along the positive orientation, draw the segments of $\partial u_1, \partial u_2$ on this annulus and check that these arcs must intersect on A, which contradicts the fact that $\partial u_1, \partial u_2$ are parallel curves on the torus T_0. Here is another way. Consider the endpoints of the edges D, X, labeled a, b, c, d on the two graphs. We

have
$$d_{v_1}(a,c) = d_{v_2}(b,d) = 1$$
so by Lemma 2.16 (applied with u_i, u_j, v_k, v_l replaced by v_1, v_2, u_1, u_2 and P, Q, R, S replaced by a, c, b, d), we should have
$$d_{u_1}(a,b) = d_{u_2}(c,d)$$
However, on Γ_a we have $d_{u_1}(a,b) = 5$ while $d_{u_2}(c,d) = 3$, which is a contradiction. □

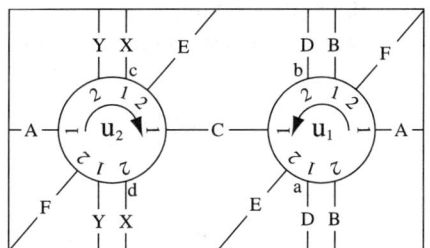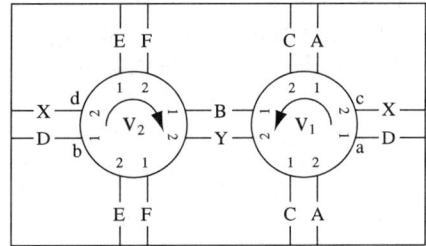

Figure 20.3

PROPOSITION 20.4. *Suppose $n_a, n_b \leq 2$. Then up to symmetry Γ_a and Γ_b are one of the following pairs.*

(1)	$-(1,1,1,1)$	$(4,0,0)$
(2)	$-(2,2,0,0)$	$(2,2,0)$
(3)	$-(2,1,1,1)$	$(3,1,1)$
(4)	$-(2,2,2,2)$	$+(4;0,0,0,0)$
(5)	$-(3,3,3,1)$	$+(3,3,3,1)$
(6)	$-(2;2,2,2,0)$	$-(3;1,1,1,1)$
(7)	$-(2;2,2,2,0)$	$-(3;2,2,0,0)$
(8)	$-(2,2,2,2)$	$-(4;0,0,0,0)$
(9)	$-(2;1,1,1,1)$	$-(2;1,1,1,1)$
(10)	$-(2;2,0,2,0)$	$-(2;2,0,2,0)$
(11)	$-(2;2,2,0,0)$	$-(2;2,2,0,0)$

PROOF. This follows from the lemmas in Sections 18–20. More precisely, the case $n_b = 1$ is done in Lemma 18.1, which gives (1)–(3) above; the case $n_a = n_b = 2$ and Γ_b positive is discussed in Lemmas 19.2–19.7 according to different numbers of loops on Γ_b, which gives (4)–(5); the case $n_a = n_b = 2$ with both graphs non-positive is discussed in Lemmas 20.2–20.3, with the possibilities listed in (6)–(11). □

PROPOSITION 20.5. *For each of the cases (3), (5), (6), (9) and (10) of Proposition 20.4, the correspondence between edges of Γ_a, Γ_b is unique up to symmetry, and is shown in Figures 18.2, 19.4, 20.4, 20.5 and 20.6, respectively.*

PROOF. For cases (3) and (5) this follows from Lemmas 18.1 and 19.7.

In case (6) we have $\Gamma_a = -(2;2,2,2,0)$ and $\Gamma_b = -(3;1,1,1,1)$. The graphs Γ_a, Γ_b are as shown in Figure 20.4. Label the edges of Γ_b as shown in the figure. By symmetry we may assume that the labels on the edge endpoints of Γ_b are as in the figure. Also up to symmetry of Γ_b on the torus $\hat{F}_b$ we may assume that the labels on v_1 are as in the figure.

The label 1 endpoints of A, B, C are non-adjacent (in the sense that one of them is not adjacent to either of the other two) among the 1-labels on ∂v_1. These are non-loops on Γ_a, and one in each family, hence their endpoints at u_1 are also non-adjacent among endpoints labeled 1. This forces the jumping number J to be ± 1. Now on Γ_a the edge A must be as shown. It is easy to see that this determines the labels on the other edges in Γ_a.

In case (9) we have $\Gamma_a = -(2; 1, 1, 1, 1)$, $\Gamma_b = -(2; 1, 1, 1, 1)$, and $\Delta = 4$, so we may assume $J = 1$. Label edge endpoints and edges of Γ_a as in the figure. Using symmetry we may assume A to be any one of the two non-loop edges labeled 1 at v_1. Then this determines the labels on the other edges. See Figure 20.5.

The determination of the edge correspondence for case (10) is similar. The graphs are shown in Figure 20.6. $\square$

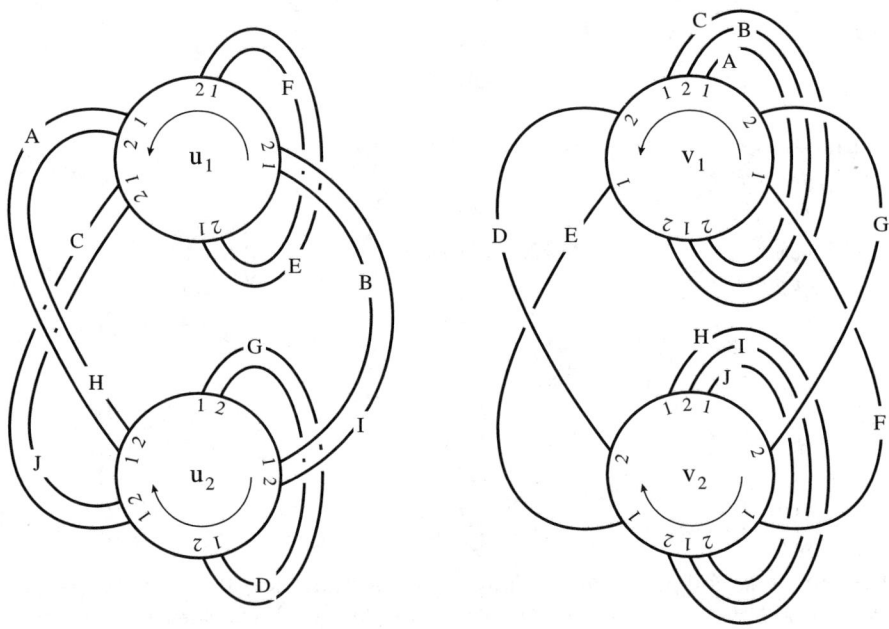

Figure 20.4

20. THE CASE $n_1 = n_2 = 2$ AND BOTH Γ_1, Γ_2 NON-POSITIVE

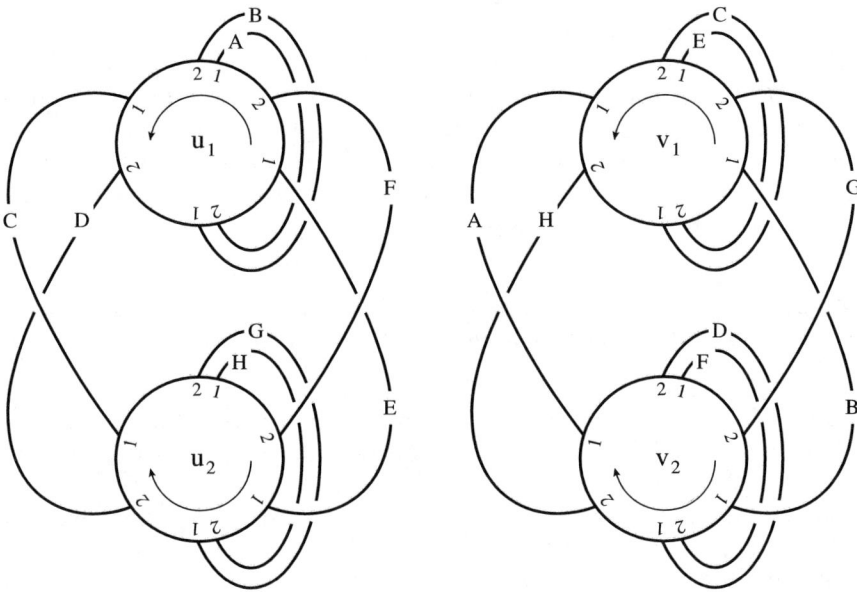

Figure 20.5

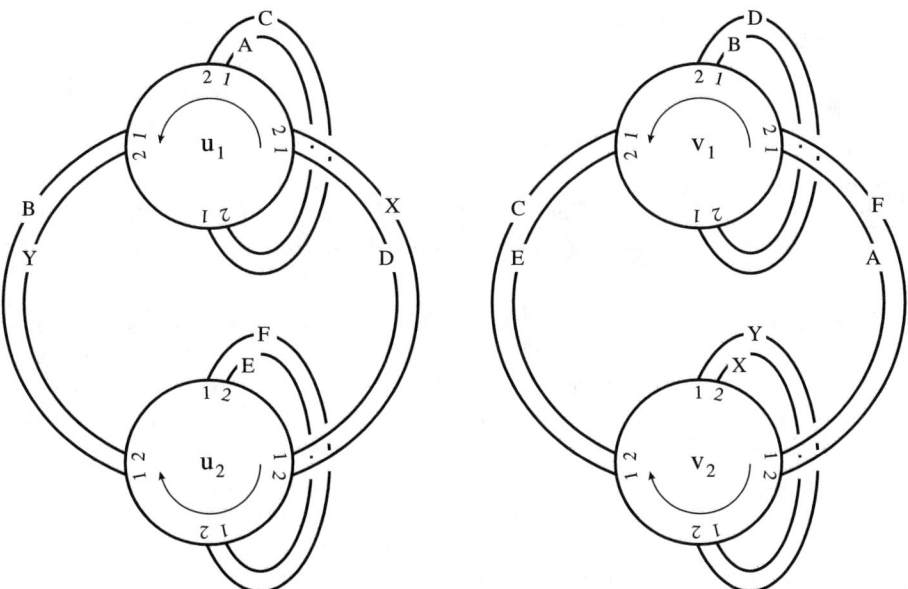

Figure 20.6

21. The main theorems

Suppose M is a hyperbolic manifold admitting two toroidal Dehn fillings $M(r_1)$, $M(r_2)$. Let F_a be essential punctured tori in M such that ∂F_a consists of a minimal number of copies of r_a, and F_1 intersects F_2 minimally. Let $X(F_1, F_2)$ be obtained from $N(F_1 \cup F_2 \cup T_0)$ by capping off its 2-sphere boundary components with 3-balls. We will use $X(r_1, r_2)$ to denote any $X(F_1, F_2)$ above with ∂F_a of slope r_a, and call it a *core* of M with respect to the toroidal slopes r_1, r_2. Note that $X(r_1, r_2)$ may not be unique.

LEMMA 21.1. *Suppose M is a hyperbolic manifold admitting two toroidal Dehn fillings $M(r_1), M(r_2)$ of distance 4 or 5. Then each of $\partial X(r_1, r_2)$ and ∂M is a union of tori.*

PROOF. By the result of the previous sections we see that Γ_a, Γ_b are either the graphs in Figures 11.9, 11.10, 14.5, 16.6, 16.8, 16.9, or one of the pairs given in Proposition 20.4.

In all figures except 11.9 and 11.10, Γ_a has two vertices and they have opposite signs. Now $X(F_1, F_2)$ can be constructed by adding thickened faces of Γ_b to $N(F_a \cup T_0)$, which has two boundary components of genus 2. It is easy to check that in all cases Γ_b has at least one disk face on each side of F_a. The boundary of a disk face of Γ_b is always an essential curve on $F_a \cup T_0$. Adding a 2-handle corresponding to a disk face will change a genus 2 boundary component to one or two tori. It follows that the boundary of $X(F_1, F_2)$ is a union of tori. Since M is irreducible and atoroidal, each torus boundary component of $X(F_1, F_2)$ either is boundary parallel, or bounds a solid torus. Therefore ∂M is also a union of tori.

The proof for Figures 11.9 and 11.10 is similar. In these cases Γ_a has 4 vertices, so $N(F_a \cup T_0)$ has two boundary components S_i of genus 3. It suffices to find two faces on each side of F_a whose boundary curves give rise to non-parallel and non-separating curves on S_i. For 11.9 one can check that the bigons on F_b bounded by the edges $K \cup J$ and $H \cup G$ are on the same side of F_a and give non-parallel boundary curves on S_1, say, while the bigon bounded by the edges $J \cup H$ and the 3-gon bounded by the edges $K \cup L \cup R$ give non-parallel non-separating curves on S_2. Hence the result follows. For 11.10, use the bigons bounded by $E \cup F$ and $G \cup H$ on one side, and the bigon $F \cup G$ and the 3-gon $E \cup K \cup A$ on the other side. □

Consider the three manifolds M_1, M_2, M_3 in [GW1, Theorem 1.1]. More explicitly, M_1 is the exterior of the Whitehead link, M_2 is the exterior of the 2-bridge link associated to the rational number $3/10$, and M_3 is the exterior of the $(-2, 3, 8)$ pretzel link, also known as the Whitehead sister link. Each of these manifolds admits two Dehn fillings $M_i(r_1)$ and $M_i(r_2)$, both toroidal and annular, with $\Delta = 4$ for $i = 1, 2$, and $\Delta = 5$ for $i = 3$. Let T_0 be the Dehn filling component of ∂M_i, and let T_1 be the other component of ∂M_i. Then for all except a few slopes s on T_1, $M' = M_i(s)$ is a hyperbolic manifold, and it admits two toroidal Dehn fillings $M'(r_1), M'(r_2)$ of distance 4 or 5. The following lemma shows that several of the cases in Proposition 20.4 can only be realized by these manifolds.

LEMMA 21.2. *Suppose Γ_a has a non-disk face. Then $M = M_i$ or $M_i(s)$ for some $i = 1, 2, 3$ and slope s on T_1, and the toroidal slopes r_1, r_2 are the same as the toroidal/annular slopes given in [GW1, Theorem 1.1].*

21. THE MAIN THEOREMS

PROOF. Let K be a curve on F_a which is essential on $\hat{F}_a$ and disjoint from $F_a \cap F_b$. Consider the manifold $X = M - \text{Int} N(K)$. If X is hyperbolic then $\hat{A} = \hat{F}_a - \text{Int} N(K)$ is an essential annulus in $X(r_a)$ and $\hat{F}_b$ is an essential torus in $X(r_b)$, so by [GW1, Theorem 1.1] $X = M_i$ for some $i = 1, 2, 3$, and we are done. Hence we may assume that X is non-hyperbolic. X is irreducible as otherwise there would be an essential sphere S in X bounding a 3-ball in M containing K, which would be a contradiction to the fact that K is an essential curve on $\hat{F}_a$. Also X cannot be a Seifert fibered manifold as otherwise $M = X \cup N(K)$ would be non-hyperbolic. Since by Lemma 21.1 ∂M is a union of tori, the above implies that X must be toroidal.

Since M is atoroidal, an essential torus T in X must be separating. Let T_0 be the Dehn filling torus component of M, and let $T_1 = \partial N(K)$. Recall that $M|T$ denotes the manifold obtained by cutting M along T. Let $V = V_T$ and $W = W_T$ be the components of $M|T$, where W is the component containing T_0. Among all essential tori in X, choose T so that (a) if there is some T in X such that $T_1 \subset W_T$, choose T so that V_T contains no essential torus; (b) if every essential torus in X separates T_0 from T_1, choose T such that W_T contains no essential torus.

Since M is atoroidal, T is inessential in M, hence V is either (i) a solid torus, or (ii) $T^2 \times I$, or (iii) a 3-ball with a knotted hole. Note that in the first two cases V must contain the curve K. Let $N = V - \text{Int} N(K)$ in the first two cases, and $N = V$ in the last case. Let $C = T \cap F_a$. Using a standard cut and past argument we may assume that each component of C is essential on both T and F_a. In case (iii) let D be a compressing disk of T in W.

Claim 1. $C \neq \emptyset$.

PROOF. If $C = \emptyset$ then F_a lies in W, which is impossible in cases (i) or (ii) because the curve K on F_a lies in V. In case (iii) $D \cap F_a$ is a set of circles and one can use the incompressibility of F_a in W to isotope F_a so that it is disjoint from D. But then D is disjoint from K, so T would be compressible in X, which is a contradiction. □

Claim 2. C is a set of essential curves on $\hat{F}_a$ parallel to K.

PROOF. Since C is disjoint from K, we need only show that each component α of C is an essential curve on $\hat{F}_a$. Assume to the contrary that α bounds a disk E on $\hat{F}_a$ and is innermost on $\hat{F}_a$. Then E must contain some boundary component of F_a, hence $E \subset W(r_a)$. In case (i) $V \cup N(E)$ is either a 3-ball, or a punctured lens space or $S^1 \times S^2$, containing the curve K, contradicting the fact that $\hat{F}_a$ is incompressible and $M(r_a)$ irreducible. In case (ii) $V \cup N(E)$ is a punctured solid torus, so the irreducibility of $M(r_a)$ implies that $M(r_a)$ is a solid torus, which is absurd because it is supposed to be toroidal. In case (iii), for homological reasons ∂E and ∂D must be homotopic on T, hence ∂E is null-homotopic in M, which contradicts the facts that C is essential on F_a and F_a is incompressible in M. □

Claim 3. Case (iii) cannot happen, i.e. V is not a 3-ball with a knotted hole.

PROOF. We have shown that all components of C are essential curves on $\hat{F}_a$ parallel to K, and $C \neq \emptyset$. Let α be a component of C. Then K is isotopic to α in $M(r_a)$, but since $\alpha \subset T$ lies in the 3-ball $V \cup N(D)$, α, and hence K, is null-homotopic in $M(r_a)$, which contradicts the fact that $\hat F_a$ is incompressible in $M(r_a)$. □

Claim 4. *W is hyperbolic.*

PROOF. Clearly W is irreducible (since X is) and not a Seifert fibered space (since M is hyperbolic). Suppose W contains an essential torus T'. By Claim 3 we see that T' cannot be of type (iii), so it must be of type (i) or (ii), which, by our choice of T, implies that every essential torus in X separates T_0 from T_1. By the choice of T, W must be atoroidal. □

We now continue with the proof of Lemma 21.2. Let A be a component of $F_a \cap W$ which contains some boundary components of F_a. By Claims 1 and 2, the corresponding component $\hat A$ of $\hat F_a \cap W(r_a)$ is an annulus in $W(r_a)$, which is incompressible because $\hat F_a$ is incompressible, and not boundary parallel because otherwise $\hat F_a$ would be isotopic to a torus with fewer intersections with the Dehn filling solid torus. Therefore $W(r_a)$ is annular.

Let P be the component of $F_a \cap V$ containing K, and let β be a component of $P \cap T$. Note that P is an annulus. Since F_b is disjoint from K, it can be isotoped to be disjoint from P, hence after isotopy we may assume that $F_b \cap T$ and $F_a \cap T$ are all parallel to β and hence mutually disjoint. If $F_b \cap T = \emptyset$ then $\hat F_b$ is an essential torus in $W(r_b)$, and if $F_b \cap T \neq \emptyset$ then as above, a component of $\hat F_b \cap W(r_b)$ which intersects the Dehn filling solid torus is an essential annulus in $W(r_b)$, hence $W(r_b)$ is either toroidal or annular. Using Theorem 1.1 of [GW1] in the first case and Theorem 1.1 of [GW3] in the second case, we see that $W = M_i$ for $i = 1$, 2, or 3.

By Claim 3 V is either a solid torus or $T^2 \times I$. In the first case $M = M_i(s)$ for some s on $T_1 = \partial V$, and in the second case $M = M_i$. □

DEFINITION 21.3. *(1) Define a set of triples (M_i, r'_i, r''_i) as follows. For $i = 1,2,3$, (M_i, r'_i, r''_i) are the manifolds and the toroidal/annular slopes given in Theorem 1.1 of [GW1]. $M_4, ..., M_{14}$ are the manifolds $X(F_1, F_2)$ corresponding to the intersection graphs given in Figures 11.9, 11.10, 14.5, 16.6, 16.8, 16.9, 18.2, 19.4, 20.4, 20.5 and 20.6, and r'_i, r''_i are the boundary slopes of the corresponding surfaces F_1, F_2.*

(2) Two triples (M, r', r'') and (N, s', s'') are equivalent, denoted by $(M, r', r'') \cong (N, s', s'')$, if there is a homeomorphism from the 3-manifold M to N which sends the boundary slopes (r', r'') to (s', s'') or (s'', s').

The following theorem shows that if a hyperbolic manifold M admits two toroidal Dehn fillings along slopes r_1, r_2 of distance 4 or 5 then (M, r_1, r_2) is either one of these triples, or obtained from such an M_i by Dehn filling on $\partial M_i - T_0$.

THEOREM 21.4. *Let M be a hyperbolic 3-manifold admitting two toroidal Dehn fillings $M(r_1), M(r_2)$ with $\Delta(r_1, r_2) = 4$ or 5. Let n_a be the minimal number of intersections between essential tori and the Dehn filling solid torus in $M(r_a)$. Assume $n_a \leq n_b$. Let (M_i, r'_i, r''_i) be the manifolds defined above, and let T_0 be the boundary component of M_i containing r'_i, r''_i. Then*

(1) $n_a \leq 2$, $n_b \leq 6$;

(2) either $(M, r_1, r_2) \cong (M_i, r'_i, r''_i)$ for some $i = 1, ..., 14$, or $(M, r_1, r_2) \cong (M_i(s), r'_i, r''_i)$, where $i \in \{1, 2, 3, 14\}$ and s is a slope on $T_1 = \partial M_i - T_0$; and

(3) $i \in \{1, 2, 4, 6, 9, 13, 14\}$ if $\Delta = 4$, and $i \in \{3, 5, 7, 8, 10, 11, 12\}$ if $\Delta = 5$.

PROOF. This is a summary of the results in the previous sections. Assume $n_a \leq n_b$. Then by Proposition 11.10 we have $n_a \leq 2$. By Proposition 16.8 if $n_b \geq 3$ then X is one of those in Figure 11.9, 11.10, 14.5, 16.6, 16.8 or 16.9.

We may now assume $n_a, n_b \leq 2$. Then by Proposition 20.4 Γ_a, Γ_b is one of the 11 pairs listed there. One can check that all but cases (3), (5), (6), (9), (10) have the property that one of $\hat{F}_a, \hat{F}_b$ contains a non-disk face, so by Lemma 21.2 the triple (M, r_1, r_2) is (M_i, r'_i, r''_i) or $(M_i(s), r'_i, r''_i)$ for some $i = 1, 2, 3$. Finally, by Proposition 20.5 the graphs of the above cases are given in Figures 18.2, 19.4, 20.4, 20.5 and 20.6.

(3) follows by counting Δ for the graph pairs of each of the manifolds listed in (2). □

22. The construction of M_i as a double branched cover

The first three of the 14 manifolds M_i have already been identified as the exteriors of links in S^3. See [GW1]. The links are shown in Figure 24.1. Besides M_4 and M_5, the other nine manifolds $M_6, ..., M_{14}$ have the property that Γ_a is a graph on $\hat{F}_a$ with two vertices of opposite signs. In this section we will construct, for each $i = 6, ..., 14$, a tangle $Q_i = (W_i, K_i)$, where W_i is a 3-ball for $i = 6, ..., 13$, and an $S^2 \times I$ for $i = 14$, such that M_i is the double branched cover of W_i with branch set K_i. It is well known that once we have such a presentation then the Dehn filling $M_i(r)$ will be the double branched cover of $Q_i(r)$, where $Q_i(r)$ is obtained by attaching a rational tangle of slope r to Q_i, with coordinates properly chosen.

Here is a sketch of the construction. Assume Γ_a is non-positive and $n_a = 2$, and suppose there is an orientation-preserving involution α_1 on F_a which maps ∂u_1 to ∂u_2 and preserves Γ_a. The restriction of α_1 on ∂F_a extends to an involution α_2 on T_0 which has four fixed points, and it preserves the curves ∂F_b on T_0. Thus $\alpha = \alpha_1 \cup \alpha_2$ is an involution on $F_a \cup T_0$, which has eight fixed points, four on each of F_a and T_0. Since α preserves $\Gamma_a \cup \partial F_b$, it extends over each disk face of F_b to give an involution on F_b. One can now further extend the involution α from $F_a \cup F_b \cup T_0$ to a regular neighborhood Y of $F_a \cup F_b \cup T_0$. For $i \geq 6$, M_i is obtained by capping off spherical boundary components of Y by 3-balls, hence α extends to an involution of M_i. Clearly the quotient of $N(F_a \cup T_0)$ is a twice punctured 3-ball W_i. After attaching 2-handles corresponding to faces of F_b and some 3-balls we see that W_i is a punctured 3-ball. From the construction below we will see that W_i is a 3-ball when $i = 6, ..., 13$, and an $S^2 \times I$ when $i = 14$. Denote by K_i the branch set of α in W_i. Then $Q_i = (W_i, K_i)$ is the tangle corresponding to the manifold M_i, and M_i is the double branched cover of Q_i in the sense that it is the double branched cover of W_i with branch set K_i. Attaching a rational tangle of slope t to T_i, we obtain a new tangle $Q_i(t)$ whose double branched cover is $M(r)$ for some slope r on $T_0 \subset \partial M$. This makes it possible to see the essential torus in $M_i(r_a)$ as a lifting of some surface in $Q_i(t)$.

To illustrate this procedure, we give below a step by step construction of the tangle $Q_6 = (W_6, K_6)$ for the manifold M_6 corresponding to the graphs in Figure 14.5. The constructions for the other manifolds are similar.

Denote by $N(C)$ a regular neighborhood of a set C in a 3-manifold, and by I the interval $[-1, 1]$.

STEP 1. *Identify* $[N(F_a \cup T_0)/\alpha] - D_2 \times I$ *with* $S^2 \times I$.

Recall that α has four branch points on each of T_0 and F_a, so $T_0/\alpha = S$ is a 2-sphere, and $F_a/\alpha = D_1$ is a disk. Let D_2 be a small disk in the interior of D_1, disjoint from Γ_b/α and the branch points of α. Then $A_1 = D_1 - \text{Int}(D_2)$ is a collar of ∂D_1. Therefore $N((F_a \cup T_0)/\alpha)$ can be written as

$$(S \times I) \cup (A_1 \times I) \cup (D_2 \times I)$$

Note that $A_1 \times I$ is a collar of the attaching annulus $\partial D_1 \times I$, hence $X = (S \times I) \cup (A_1 \times I)$ is homeomorphic to $S^2 \times I$.

One boundary component of $X = S^2 \times I$ is $\partial_- X = (T_0 \times \{-1\})/\alpha$, and the other boundary component $\partial_+ X$ can be written as $D_+ \cup D_- \cup A$, where the two disks $D_+ \cup D_- = \partial X \cap (S \times 1)$ lift to two annuli on $T_0 \times 1$, and A is the annulus $\partial X \cap (A_1 \times I)$. We identify X with $(\mathbb{R}^2 \cup \{\infty\}) \times I$, so that the disks $D_\pm$ are identified with the squares $I \times [\pm 2, \pm 4]$ on the plane $P = \mathbb{R}^2 \times 1$ on ∂X, the annulus A is the closure of $P - D_+ \cup D_-$, and the core c_0 of A is identified with the closure of the x-axis of P. See Figure 22.1. (Not drawn to scale.)

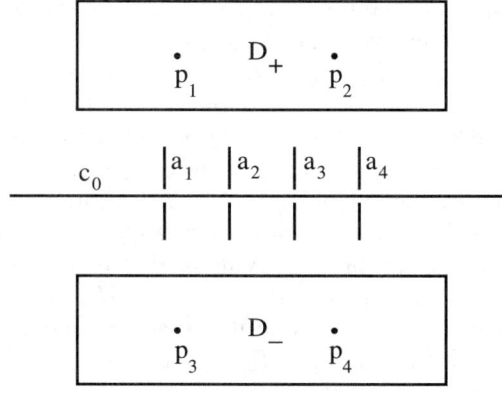

Figure 22.1

The branch set of α now consists of eight arcs. Four of them come from the fixed points of α on $T_0 \times I$, and are of type $p_i \times I$ in X, where $p_1, p_2 \in D_+$, and $p_3, p_4 \in D_-$. These will be represented by four dots $p_1, ..., p_4$ on P, two in each of $D_\pm$, as shown in Figure 22.1. The other four branch arcs of α are of type $a_i = q_i \times I \subset A \times I$, where $q_1, ..., q_4$ are the branch points of α in $A = D_1 - \text{Int}(D_2)$. Note that a_i has both endpoints on $S^2 \times 1$. We may assume that these project to four vertical arcs on the annulus A in P above, and we may arrange so that the endpoints of these arcs have y-coordinates ± 1 on the plane P. See Figure 22.1. Denote by $a_i(1)$ and $a_i(-1)$ the endpoints of a_i with y coordinates 1 and -1, respectively.

STEP 2. *Draw the arcs $G' = (\Gamma_a \times \{\pm 1\})/\alpha$ on P, with edges and edge endpoints labeled.*

The graph $\Gamma_a \times 1$ on $F_a \times 1$ projects to a set of arcs E on $D_1 \times 1$. We may choose the disk D_2 above to be disjoint from E. Then E lies in the annulus $A_+ = A \times 1$. If a family $\hat{e}_i$ has $2k$ edges then they project to k edges on A_+ with endpoints on ∂D_+, each circling around the branch point $a_i(1)$. If $\hat{e}_i$ has $2k + 1$ edges then the quotient is a set of k edges as above together with an edge connecting a point on

22. THE CONSTRUCTION OF M_i AS A DOUBLE BRANCHED COVER

∂D_+ to $a_i(1)$. Up to isotopy we may assume that all edge endpoints of E on ∂D_+ lie on the horizontal line $y = 2$ on P. Similarly the projection of $\Gamma_a \times (-1)$ is a set of arcs on the annulus $A_- = A \times (-1)$, which is the mirror image of the arcs $(\Gamma_a \times 1)/\alpha$ along the circle c_0 on P. Denote by G' the set of arcs above.

For the graph Γ_a in Figure 14.5(a), the edges in G' are shown in Figure 22.2. The edges are labeled by the corresponding edges in Γ_a. (We only show a few of the labels in the figure; the others should be easy to identify.) Each edge in G' is the image of two edges in Γ_a, hence it has two labels. (Note that if one of the families has an odd number of edges then the middle one projects to an arc in G' with a single label.) All arcs appear in the region $I \times [-2, 2]$. The top and bottom lines in the figure represent arcs on $\partial D_\pm$. Note that each edge endpoint on ∂D_+ corresponds to one edge endpoint on each of ∂u_1 and ∂u_2. The labels on the top and bottom lines correspond to the labels on ∂u_1 in Figure 14.5(a).

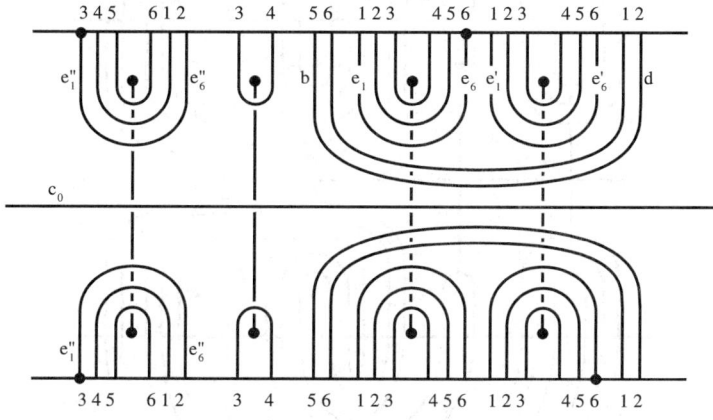

Figure 22.2

STEP 3. *Add arcs $G'' = (\partial F_b \times 1)/\alpha$ on P to obtain $G = G' \cup G''$.*

Recall that the preimage of $D_\pm$ are two annuli $A_\pm$ on $T_0 \times 1$. The curves $G'' = (\partial F_b \times 1 \cap A_\pm)/\alpha$ is now a set of arcs in $D_\pm$, with $\partial G''$ the union of $\partial G'$ and possibly some of the branch points p_i in $D_\pm$. We need to determine how the endpoints of G' are connected by the edges of G''.

Consider the circle ∂v_6 in Figure 14.5(b). We may assume that the segments on ∂v_6 from label 1 to label 2 (in the counterclockwise direction) project to arcs in D_+ while those from label 2 to label 1 project to arcs in D_-. Consider the arc β on ∂v_6 from the tail of e_6 to the head of e_6''. (Recall that e_6'' is the edge between e_6 and e_6' in Figure 14.5(a)). Note that the tail of e_6 projects to the endpoint of $e_1 = e_6$ with label 6 on ∂D_+ in Figure 22.2. The other endpoint q of β is the head of e_6'', which lies on ∂u_2. Since the labels in Figure 22.2 are the ones corresponding to those on ∂u_1 in Figure 14.5(a), we have to find the corresponding point on ∂u_1 in order to determine the position of the edge endpoint q on ∂D_+. On Figure 14.5(a) the involution α restricted to ∂u_2 is a vertical translation, which maps the head of e_6'' (i.e. the edge in $\hat{e}_3$ labeled 6 at u_2) to the tail of e_1'', which has label 3 at u_1. It follows that q is the endpoint of $e_1'' = e_6''$ in Figure 22.2 with label 3 at ∂D_+. The two endpoints of β are represented by the two dots on the top line in Figure 22.2.

Similarly, let β' be the arc on ∂v_6 from the head of e_6'' to the tail of e_6'. Then it is an arc in D_- with endpoints on the dots at the bottom line in Figure 22.2.

The arcs G'' in $D_\pm$ are parallel to each other, and they are non-trivial in the sense that none of them cuts off a disk in $D_\pm$ that does not contain a branch point of α. Therefore the above information completely determines the arcs G'' as well as the branch points $p_1, ..., p_4$ of α. (Note that if the number of edge endpoints between the dots is odd then the middle arc will have an endpoint on a branch point p_i.) The graph $G = G' \cup G''$ is now shown in Figure 22.3.

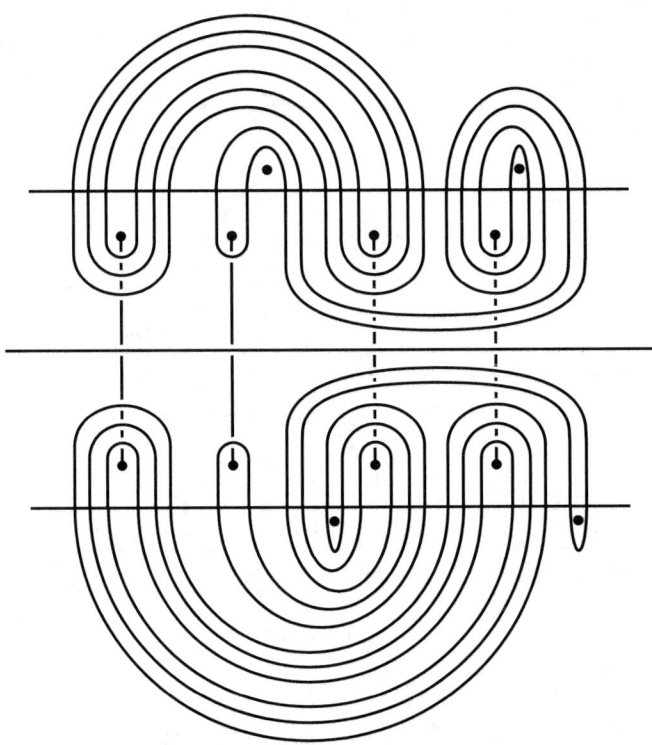

Figure 22.3

STEP 4. *Construct the tangle Q_i.*

Each component $c \neq c_0$ of G lifts to curves on F_b bounding disk faces σ of Γ_b. The quotient of $\sigma \times I$ is either a 2-handle attached to c if c is a circle, or a 3-ball attached to a neighborhood of c if c is an arc. Examining the branch set of α in $\sigma \times I$ gives the following procedure. We use X to denote the initial manifold at the beginning of each step below. In particular, $X = S^2 \times I$ before the first step.

(1) If c is an arc then $X \cup (\sigma \times I)/\alpha$ is homeomorphic to X. The new branch set is obtained by adding a trivial arc in $(\sigma \times I)/\alpha$ joining the two endpoints of c. Therefore we can simply modify the branch set of α by pushing c into the interior of X.

(2) If $c \neq c_0$ is a circle component bounding a disk D_1 on ∂X containing no branch point of α, then attaching a 2-handle along c creates a 2-sphere boundary component, which must bound a 3-ball in M_i/σ. Thus after attaching the 2-handle

and the 3-ball the manifold is homeomorphic to X, and the homeomorphism maps the new branch set to the old one. Therefore in this case we can simply delete the curve c from G.

(3) If $c \neq c_0$ is a circle component bounding a disk D_1 on ∂X containing one branch point of α, then c lifts to a circle on the boundary of a face σ of Γ_b, which necessarily contains a fixed point of α. Hence the cocore of the corresponding 2-handle is a branch arc of α. The 2-sphere boundary component created after attaching the 2-handle contains two branch points of α, hence bounds a 3-ball containing a trivial arc as branch set of α. Thus after attaching the 2-handle and the 3-ball the manifold is homeomorphic to X, and the branch set of α has not changed. As in Case (2), we will simply delete the curve c from G in this case.

(4) If a circle component $c \neq c_0$ of G bounds a disk D_1 containing exactly two branch points of α, then after attaching a 2-handle and a 3-cell, the manifold is homeomorphic to X, and the branch set of α is obtained by adding a trivial arc in the 3-cell joining the two branch points of α in D_1. Therefore in this case we will add an arc in D_1 joining the two branch points of α, push the arc into the interior of X as branch set of α, and then delete the curve c.

(5) If c is a circle component of G bounding a disk D_1 containing $k > 2$ branch points of α, simply attach a 2-handle along c. If k is odd, add an arc in the center of the 2-handle to the branch set of α.

(6) Finally, attach a 2-handle along c_0, fill each 2-sphere boundary component containing at most 2 branch points with a 3-ball, and add a trivial arc in the 3-ball to the branch set if the 2-sphere contains exactly two branch points. If the 2-sphere contains four branch point, shrink it by an isotopy to a small sphere, which projects to a small disk on the diagram, with four branch arcs attached. (This happens only for Q_{14}. See Figure 22.13.) This completes the construction of the tangle Q_i.

For M_6, the above procedure produces the tangle $Q_6 = (W_6, K_6)$ in Figure 22.4(a), where K_6 should be considered as a tangle lying in the half space Q_6 (including ∞) in front of the blackboard. The four boundary points of K_6 lie on the blackboard, which is the boundary of W_6.

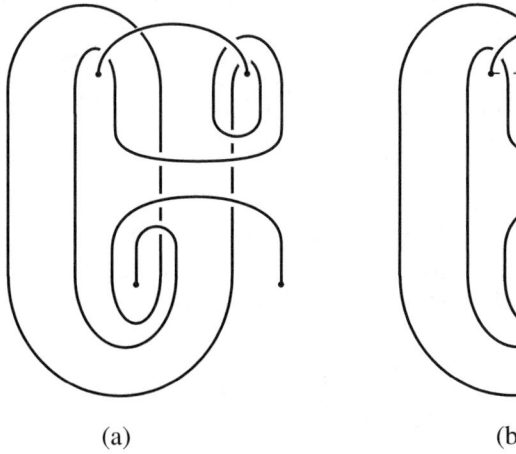

Figure 22.4

STEP 5. *Find the tangles* $Q_i(t_a) = M_i(r_a)/\alpha$.

There is one branch point p_i in each quadrant of $P' = \mathbb{R}^2$. Let m, l be curves on T_0 that project to the y-axis union ∞ and the x-axis union ∞ on P', respectively. This sets up coordinate systems on T_0 and P'. For $t = p/q$ a rational number or ∞, let $M_i(t)$ denote the Dehn filling along slope $pm + ql$, and $Q_i(t)$ denote the tangle obtained by attaching a rational tangle of slope t to P'. In other words, $Q_i(t)$ is obtained by attaching a 3-ball to Q_i on P', and adding two arcs on P' connecting the branch points of α, which lift to curves of slope t on T_0. Since the attached rational tangle lifts to a solid torus with meridional slope t on T_0, $M_i(t)$ is the double branched cover of $Q_i(t)$.

By construction ∂F_a projects to the x-axis, hence $M_i(r_1) = M_i(0)$. The slope r_2 can be obtained by connecting the curves G'' in $D_\pm$ by vertical arcs in $A = P' - \cup D_\pm$. For M_6, one can check that the slope $r_2 = 4$.

Denote by $T(a_1, a_2)$ a Montesinos tangle which is the sum of two rational tangles of slopes $1/a_1$ and $1/a_2$, respectively, where a_1, a_2 are integers. Denote by $T(a_1, b_1; a_2, b_2)$ the collection of pairs (S^3, L) which can be obtained by gluing two tangles $T(a_i, b_i)$ along their boundary. Denote by $X(a_1, a_2)$ the collection of Seifert fiber spaces with orbifold a disk with two cone points c_1, c_2 of index a_1 and a_2, i.e. the cone angle at c_i is $2\pi/a_i$. Note that the double branched cover of $T(a_1, a_2)$ is in $X(a_1, a_2)$. Denote by $X(a_1, b_1; a_2, b_2)$ the collection of graph manifolds which are the union of two manifolds X_1, X_2 glued along their boundary, where $X_i \in X(a_i, b_i)$.

Denote by $K_{p/q}$ the two bridge knot or link associated to the rational number p/q. Denote by $C(p_1, q_1; p_2, q_2)$ the link obtained by replacing each component K_i of a Hopf link by its (p_i, q_i) cable K_i', where q_i is the number of times K_i' winds around K_i. Denote by $Y(p_1, q_1; p_2, q_2)$ the double branched cover of S^3 with branch set $C(p_1, q_1; p_2, q_2)$. Denote by $C(C; p, q)$ the link obtained by replacing one component K_1 of a Hopf link by a Whitehead knot in the solid torus $N(K_1)$, and the other component K_2 by a (p, q) cable of K_2. Let $Y(C; p, q)$ be the double branched cover of S^3 with branch set $C(C; p, q)$. Denote by Z the double branched cover of S^3 with branch set the 2-string cable of the trefoil knot shown in Figure 22.13(d).

If $Q_i(r) = (S^3, L)$ then we will sometimes simply write $Q_i(r) = L$.

LEMMA 22.1. *(1)* $Q_6(0) \in T(2, 6; 2, 3)$, *as shown in Figure 22.4(b).*
(2) $Q_6(4) = C(3, 1; 2, 5)$, *as shown in Figure 22.5(b).*
(3) $Q_6(\infty) = K_{2/9}$.

PROOF. (1) The tangle $Q_6(0) = (S^3, L)$ is shown in Figure 22.4(b). A horizontal line at the middle of the diagram corresponds to a 2-sphere S which cuts the link L into two Montesinos tangles $T(2, 6)$ and $T(2, 3)$.

(2) The tangle $Q_6(4)$ is shown in Figure 22.5(a), which can be isotoped to that in Figure 22.5(b). One can see that it is the link $C(3, 1; 2, 5)$ in S^3.

(3) The tangle $Q_6(\infty)$ is shown in Figure 22.5(c). One can check that it is isotopic to the knot $K_{2/9}$ in Figure 22.5(d). □

22. THE CONSTRUCTION OF M_i AS A DOUBLE BRANCHED COVER

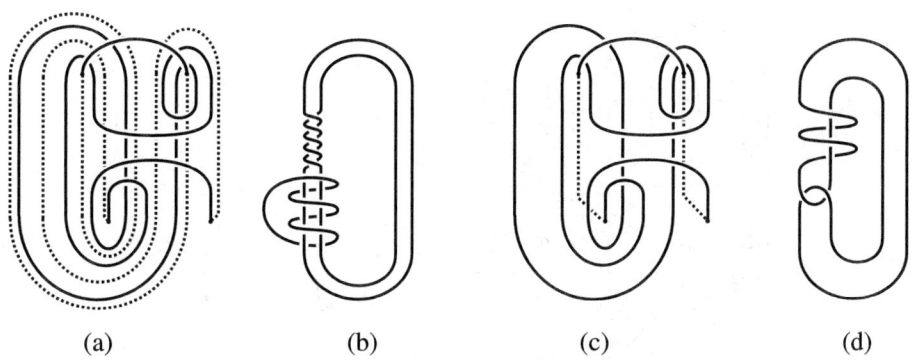

(a) (b) (c) (d)

Figure 22.5

LEMMA 22.2. *(1) For $i = 6, ..., 14$, each M_i is the double branched cover of a tangle $Q_i = (W_i, K_i)$, where Q_i is shown in Figure 22.4(a) for $i = 6$, and in Figure 22.i(b) (with dotted lines removed) when $i > 6$.*

(2) Each M_i ($i = 6, ..., 13$) admits a lens space surgery $M_i(r_3)$. For each i, let r_1, r_2 be the slopes r'_i, r''_i given in Definition 21.3. Then the manifolds $M_i(r_1)$, $M_i(r_2)$ and $M_i(r_3)$ are given in the following table.

$M_6(0) \in X(2, 6; 2, 3)$ $M_6(4) = Y(3, 1; 5, 2)$ $M_6(\infty) = L(9, 2)$
$M_7(0) \in X(2, 3; 3, 3)$ $M_7(-5/2) \in X(2, 3; 2, 2)$ $M_7(\infty) = L(20, 9)$
$M_8(0) \in X(2, 2; 2, 6)$ $M_8(-5/4) = Y(3, 1; 2, 5)$ $M_8(-1) = L(4, 1)$
$M_9(0) \in X(2, 4; 2, 4)$ $M_9(-4/3) = Y(3, 1; 2, 4)$ $M_9(-1) = L(8, 3)$
$M_{10}(0) \in X(2, 3; 2, 3)$ $M_{10}(-5/2) = Y(C; 2, 1)$ $M_{10}(\infty) = L(14, 3)$
$M_{11}(0) \in X(2, 4; 2, 4)$ $M_{11}(-5/2) = Y(C; 2, 1)$ $M_{11}(\infty) = L(24, 5)$
$M_{12}(0) \in X(2, 3; 2, 3)$ $M_{12}(5) = Y(3, 1; 2, 3)$ $M_{12}(\infty) = L(3, 1)$
$M_{13}(0) \in X(2, 3; 2, 3)$ $M_{13}(4) = Z$ $M_{13}(\infty) = L(4, 1)$

PROOF. The result for M_6 follows from Lemma 22.1 because $M_6(r)$ is a branched cover of $Q_6(r)$. The proof for the other cases are similar. Each $M_i(r)$ is the double branched cover of $Q_i(r)$ and the tangle $Q_i(r)$ is a link L in S^3. More explicitly, Figure 22.i(a) shows the curves $G = G' \cup G''$ in Step 3 of the above construction; Figure 22.i(b) gives the tangle Q_i as well as $Q_i(r_1)$, which is obtained by attaching a 0-tangle (the two horizontal dotted lines) to Q_i; Figure 22.i(c) gives $Q_i(r_2)$, which is simplified to that in Figure 22.i(d); $Q_i(r_3)$ is in Figure 22.i(e), which is simplified to that in Figure 22.i(f) for some i. (The figures are numbered so that Figure 22.i corresponds to the manifold M_i for $i \geq 7$. Note that there is no Figure 22.6.) The manifold $M_{14}(r_3)$ is the double branched cover of $Q_{14}(r_3) = T(2, 2)$ in Figure 22.14(e), and hence is a twisted I-bundle over the Klein bottle. □

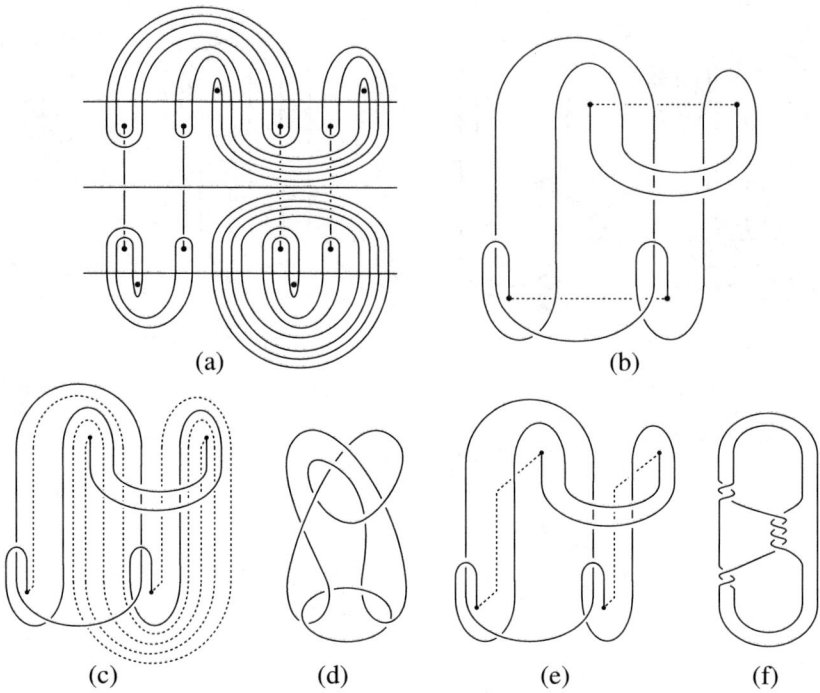

Figure 22.7

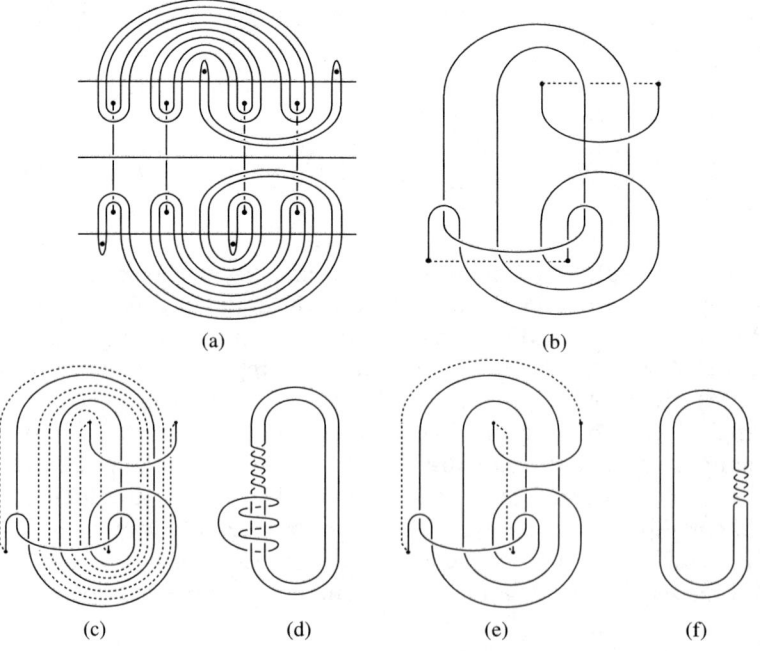

Figure 22.8

22. THE CONSTRUCTION OF M_i AS A DOUBLE BRANCHED COVER 119

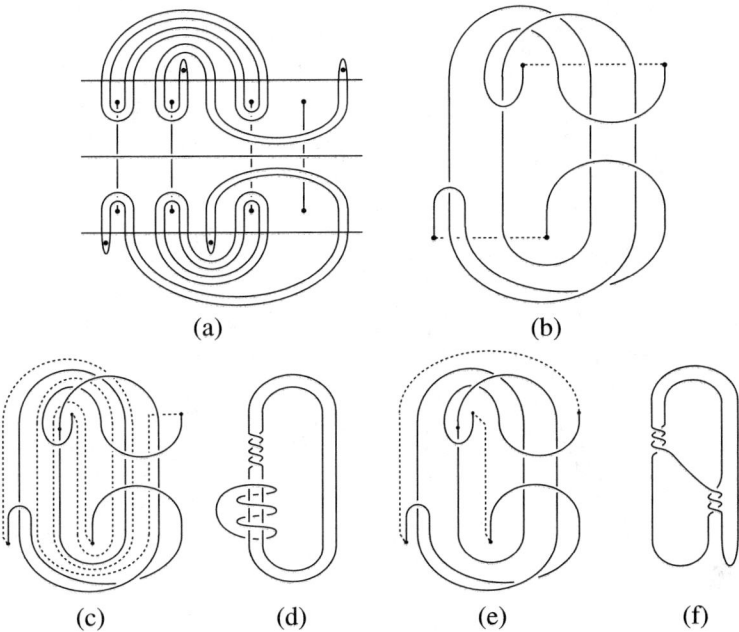

Figure 22.9

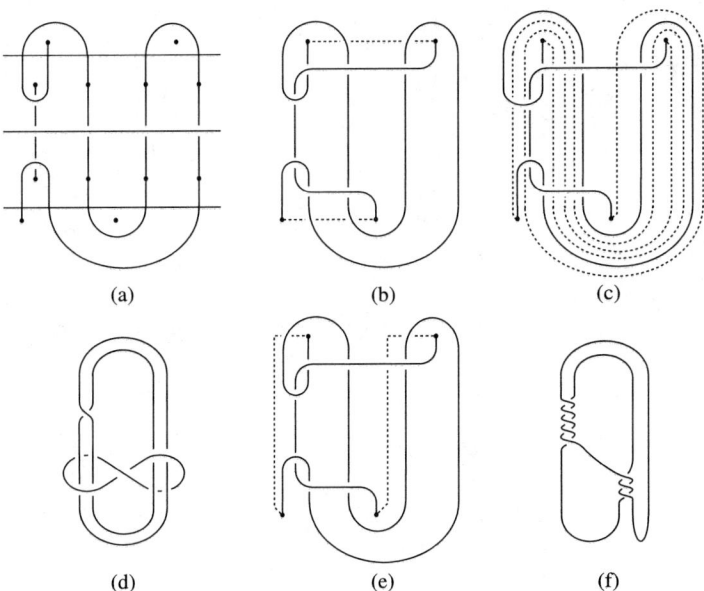

Figure 22.10

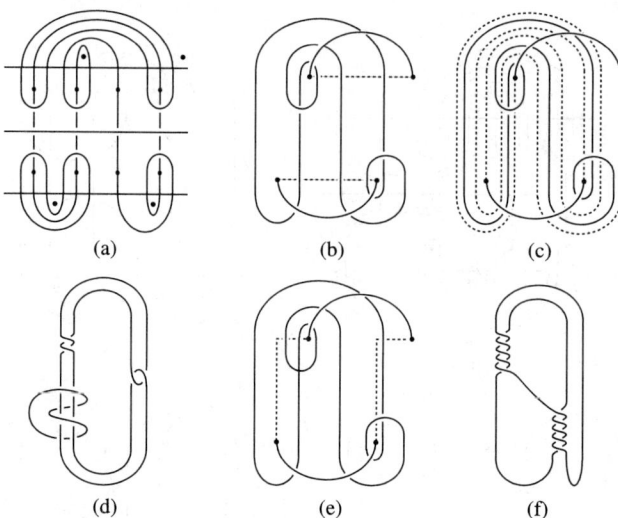

Figure 22.11

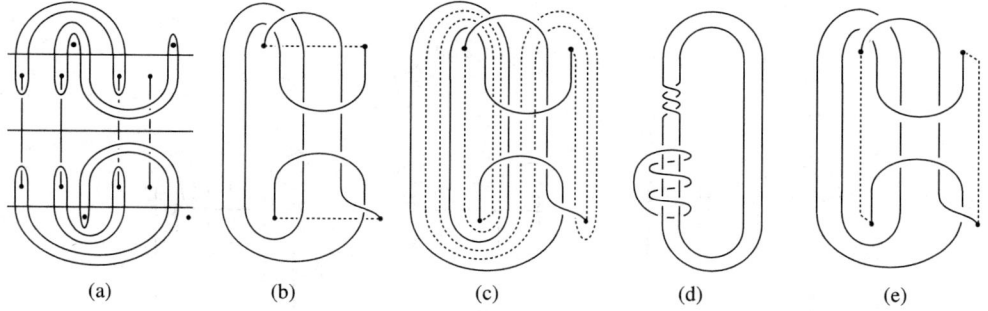

Figure 22.12

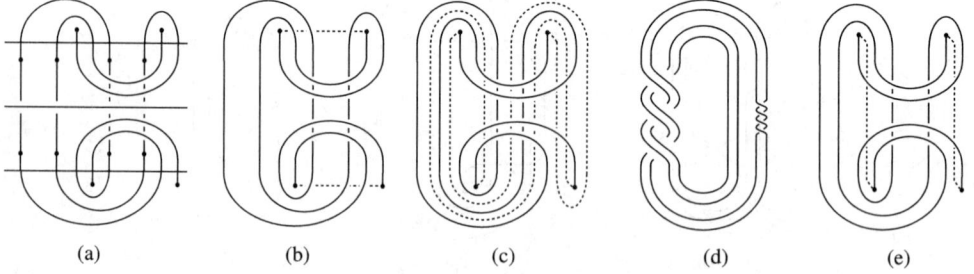

Figure 22.13

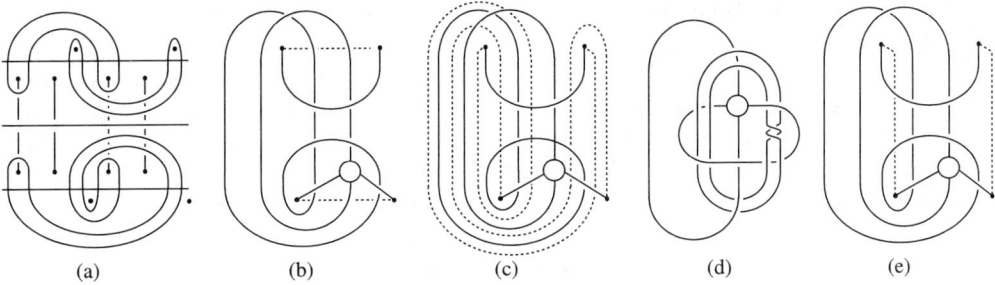

Figure 22.14

Recall that a manifold M with a fixed torus $T_0 \subset \partial M$ is large if $H_2(M, \partial M - T_0) \neq 0$. Teragaito [T2] proved that there is no large hyperbolic manifold M admitting two toroidal fillings of distance at least 5. The following result shows that there is only one such manifold for $\Delta = 4$.

THEOREM 22.3. *Suppose (M, T_0) is a large manifold and M is hyperbolic and contains two toroidal slopes r_1, r_2 on T_0 with $\Delta(r_1, r_2) \geq 4$. Then M is the Whitehead link exterior, and $\Delta(r_1, r_2) = 4$.*

PROOF. Let r be a slope on T_0, V_r the Dehn filling solid torus in $M(r)$, and K_r the core of V_r. By duality we have $H_2(M, \partial M - T_0) \cong H^1(M, T_0)$, which is isomorphic to the free part of $H_1(M, T_0)$. Also,

$$H_1(M, T_0) \cong H_1(M(r), V_r) \cong H_1(M(r))/H_1(K_r).$$

Put $G(M, r) = H_1(M(r))/H_1(K_r)$. Then we need only show that $G(M_i, r)$ is a (possibly trivial) torsion group for $i = 2, ..., 14$ and r some slope on T_0.

For $i = 2, 3$, M_i is the exterior of a closed braid K_i in a solid torus V. Let r be the meridian slope of K_i. Then $G(M_i, r) = \mathbb{Z}_p$, where p is the winding number of K_i in V.

For $i = 6, ..., 13$, by Lemma 22.2 M_i has a lens space filling $M_i(r_3)$. Therefore $G(M_i, r_3)$ is a quotient of the finite cyclic group $H_1(M_i(r_3))$ and hence is a torsion group. Similarly for the four manifolds in [Go] with toroidal slopes of distance at least 6.

For $i = 14$, take a regular neighborhood of $u_1 \cup u_2 \cup D$ on $\hat{F}_a$ as a base point. See Figure 20.6(a). Then $H_1(M_{14}(r_a))$ is generated by x, y, s_1, s_2, where x is the element of $H_1(\hat{F}_a)$ represented by the edges C on Figure 20.6(a), oriented from the label 2 endpoint to the label 1 endpoint, y is represented by B, oriented from u_1 to u_2, and s_i by the part of the core of the Dehn filling solid torus running from u_i to u_{i+1} with respect to the orientation of ∂F_b. Then the bigons $B \cup D$, $C \cup E$ and the 4-gon bounded by $C \cup D \cup E \cup Y$ on F_b give relations $2s_1 - y = 0$, $2x = 0$, and $y + 2x = 0$. The other faces of Γ_b are parallel to these. To calculate $G(M_{14}, r_a) = H_1(M_{14}(r_a))/H_1(K_a)$ we further add the relation $s_1 + s_2 = 0$. One can now check that $G(M_{14}, r_a) = \mathbb{Z}_2 \oplus \mathbb{Z}_2$, and the result follows.

For $i = 4$, choose a regular neighborhood of $v_1 \cup v_2 \cup J$ in Figure 11.9(b) as a base point. Then $H_1(M_4(r_b))$ is generated by x, y, s_1, s_2, where x, y are represented by the edges L, C in Figure 11.9(b), oriented from v_1 to v_2, and s_i by the part of the core of the Dehn filling solid torus from v_i to v_{i+1}. The faces bounded by $L \cup C$, $C \cup K$ and $Q \cup K \cup M \cup A$ give the relations $y - s_1 + x + s_2 = 0$, $s_1 - x - s_2 = 0$, and $s_2 - s_1 + y = 0$. Together with the relation $s_1 + s_2 = 0$ from $H_1(K_b) = 0$, these give $G(M_4, r_b) = \mathbb{Z}_2$.

For $i = 5$, $H_1(M_5(r_b))$ is generated by x, y, s, where x, y are represented by edges E and C on Figure 11.10(b), oriented from label 3 to label 4, and s is represented by the core of the Dehn filling solid torus. Then the bigon $A \cup H$ and the annulus bounded by $A \cup G \cup C \cup E$ on Figure 11.10(a) containing J give the relations $x + y = 0$ and $2x - 2y = 0$. Adding the relation $s = 0$ gives $G(M_5, r_b) = \mathbb{Z}_4$. □

23. The manifolds M_i are hyperbolic

The manifolds M_1, M_2, M_3 in Definition 21.3 are known to be hyperbolic, see [GW1, Theorem 1.1]. In this section we will show that the other 11 manifolds M_i in Definition 21.3 are also hyperbolic. See Theorem 23.14 below.

A knot K in a solid torus V is a (p, q) knot if it is isotopic to a (p, q) curve on ∂V with respect to some longitude-meridian pair on ∂V. In particular, the winding number of K in V is p.

LEMMA 23.1. *(1) If $i \in \{6, ..., 13\}$ and $j = 1$ or $i \in \{6, 7, 8, 9, 12, 13\}$ and $j = 2$, then $M_i(r_j)$ contains an essential torus T cutting it into two Seifert fiber spaces E_1, E_2.*

(2) For $i = 10, 11$, $M_i(r_2)$ contains a non-separating essential torus cutting $M_i(r_2)$ into a Seifert fiber space whose orbifold is an annulus with a cone point of index 2.

(3) For $i = 6, ..., 13$ and $j = 1, 2$, $M_i(r_j)$ is irreducible, and contains no hyperbolic submanifold bounded by an incompressible torus.

PROOF. (1) By Lemma 22.2, for $i = 6, ..., 13$, $M_i(r_1)$ is of type $X(a_1, b_1; a_2, b_2)$, which is the union of two Seifert fiber spaces of types $X(a_1, b_a)$ and $X(a_2, b_2)$, hence the result is true for $M_i(r_1)$. Similarly it is true for $M_7(r_2)$.

For $i = 13$, $Q_i(r_2) = (S^3, L)$, where L is the link in Figure 22.13(d), which consists of two parallel copies of the trefoil knot. The two components of L bound an annulus A. Cutting S^3 along A gives the trefoil knot exterior E, and A becomes a torus T. The double branched cover of $Q_{13}(r_2)$ is obtained by gluing two copies of E along T. Hence the result is true because E is a Seifert fiber space and T is incompressible in E.

For $i = 6, 8, 9, 12$, $Q_i(r_2) = (S^3, L)$ is of type $C(p_1, q_1; p_2, q_2)$, so there is a torus T' cutting S^3 into two solid tori V_1, V_2, such that each $L_j = L \cap V_j$ is a (p_j, q_j) knot in V_j for some $p_j > 1$. Note also that in these cases at least one of the p_j is odd, which implies that T' lifts to a single torus T, cutting $M_i(r_2)$ into two components W_1, W_2, such that W_j is a double branched cover of (V_j, L_j). The (p_j, q_j) fibration of V_j now lifts to a Seifert fibration of W_j, hence the result follows.

(2) For $i = 10, 11$, $Q_i(r_2) = (S^3, L)$, and there is a torus T' cutting S^3 into two solid tori V_1, V_2, such that $L_1 = V_1 \cap L$ is a $(2, 1)$ knot, and $L_2 = V_2 \cap L$ is a Whitehead knot in the solid torus V_2. Since both winding numbers of L_j are even, T' lifts to two tori in $M_i(r_2)$. Let W_i be the lifting of V_i. A meridian disk of V_1 lifts to an annulus in W_1, hence W_1 is a $T^2 \times I$ (not a twisted I-bundle over the Klein bottle because ∂W_1 has two components). Let T be the core of this $T^2 \times I$. Then it cuts $M_i(r_2)$ into the manifold W_2. We need to show that W_2 is Seifert fibered.

Let D be a meridian disk of V_2 which intersects L_2 at two points. Then $(V_2, L_2) = (B_1, L'_1) \cup (B_2, L'_2)$, where $B_1 = N(D)$, B_2 is the closure of $V_2 - B_1$, and $L'_k = L_2 \cap B_k$. Note that each L'_k is a trivial tangle in B_k, hence its double branched cover V'_k is a solid torus. One can check that each component of $V'_1 \cap V'_2$ is a longitudinal annulus on $\partial V'_1$, and it is an annulus on $\partial V'_2$ with winding number 2 in V'_2. Therefore $W_2 = V'_1 \cup V'_2$ is a Seifert fiber space whose orbifold is the union of a D^2 and a $D^2(2)$ glued along two boundary arcs. Since W_2 has two torus boundary components, the orbifold must be an annulus with a single cone point of index 2.

(3) Let T be the essential torus in $M_i(r_j)$ given in the above proof. Then it cuts $M_i(r_j)$ into one or two bounded Seifert fiber spaces, which are irreducible. Since T is incompressible, $M_i(r_j)$ is also irreducible. The second statement follows from the fact that the JSJ (Jaco-Shalen-Johannson) decomposition of an irreducible closed 3-manifold is unique. $\square$

A (p', q') knot K in a solid torus V is also called a 0-bridge knot. In this case there is an essential annulus in $V - \text{Int} N(K)$ with one boundary component in each of ∂V and $\partial N(K)$. This defines a longitude l for K, which is unique if K is not the core of V. A (p, q) cable of a 0-bridge knot K is a knot on $\partial N(K)$ which represents $pl + qm$ in $H_1(\partial N(K))$, where m is a meridian of K. We refer the reader to [Ga1] for the definition of a 1-bridge braid in V.

LEMMA 23.2. *Suppose X is an irreducible, ∂-irreducible, compact, orientable 3-manifold with $\partial X = T_1 \cup T_0$ a pair of tori. Let r_1, r_2 be distinct slopes on T_0 such that $X(r_1), X(r_2)$ are both ∂-reducible. Let K_a be the core of the Dehn filling solid torus in $X(r_a)$. Then one of the following holds, up to relabeling of r_i.*

(1) Each $X(r_a)$ is a solid torus, K_a is a 0- or 1-bridge braid in $X(r_a)$, and $\Delta(r_1, r_2) = 1$ if it is not a 0-bridge knot.

(2) $X(r_1)$ is a solid torus, and $X(r_2) = (S^1 \times D^2) \# L(p, q)$ with $p \geq 2$. K_1 is a (p, q) cable of a (p', q') knot in $X(r_1)$, and r_2 is the cabling slope of K_1 in $X(r_1)$. Moreover, if m_a is the slope on T_1 bounding a disk in $X(r_a)$, then $\Delta(m_1, m_2) = pp'$.

PROOF. If both $X(r_1), X(r_2)$ are irreducible then they are solid tori and (1) holds by [Ga1, Theorem 1.1] and [Ga2, Lemma 3.2]. Now assume $X(r_2)$ is reducible. Then by [Sch, Theorem 6.1] K_1 is a (p, q) cable of some knot K' in $X(r_1)$ with respect to some meridian-longitude pair (m, l) of K', and r_2 is the cabling slope. In this case $X(r_2)$ is a connected sum $W_1 \cup L(p, q)$, where W_1 is obtained by surgery on the knot K' in $X(r_a)$ along the cabling slope $r' = pl + qm$. Denote by m' the meridian slope of K'. Then $\Delta(m', r') = p > 1$.

Denote by $K'(s)$ the manifold obtained by s-surgery on K' in $V = X(r_1)$. The assumption on X implies that T_1 is incompressible in $V - K'$, and $V - K'$ is irreducible. By the above, T_1 is compressible in both $K'(m') = V$ and $K'(r') = W_1$, and $\Delta(m', r') > 1$, hence by [Wu2, Theorem 1.1] and [CGLS, Theorem 2.4.3], either $V - \text{Int} N(K') = T^2 \times I$, or there is an annulus A in $V - \text{Int} N(K')$ with one boundary component on T_1 and another boundary component a curve of slope r on $T' = \partial N(K')$, satisfying $\Delta(r, m') = \Delta(r, r') = 1$. In either case K' is isotopic to a curve on T_1 and hence is a 0-bridge knot. Since $V - K$ is irreducible, this implies that V is also irreducible. Therefore $V = X(r_1)$ is a solid torus, K' is a (p', q') knot in V for some (p', q'), and r is the cabling slope of K' when $p' > 1$.

If $p' = 1$, i.e. $V - \text{Int} N(K') = T^2 \times I$, then K is the (p, q) cable of the core K' of V, and r_2 is the cabling slope of K. We have $X(r_2) = V \# L(p, q)$, and the slope m_2 on T_1 which bounds a disk in $X(r_2)$ is the (p,q) curve on T_1, hence $\Delta(m_1, m_2) = p = pp'$.

Now assume $p' > 1$. Choose the longitude l of K' to be the cabling slope r of K' given above. Since $r' = pr + qm$, the equation $\Delta(r, r') = 1$ above implies that $q = \pm 1$. Reversing the orientation of m' if necessary we may assume $q = 1$. Hence K is a $(p, 1)$ cable of a (p', q') knot in V, and r_2 is the cabling slope. It is easy to see that the meridian slopes m_a of $X(r_a)$ satisfy $\Delta(m_1, m_2) = pp'$, and the result follows. □

LEMMA 23.3. *Let $i = 6, ..., 13$. Let α be the covering transformation of the double branched cover $M_i \to Q_i$.*

(1) M_i is irreducible, not Seifert fibered, and contains no non-separating torus.

(2) If M_i is not hyperbolic then it contains a separating essential torus T such that T is α-equivariant, and the component W of $M_i|T$ which does not contain T_0 is either Seifert fibered or hyperbolic.

PROOF. (1) If M_i is reducible then the summand which does not contain T_0 is a summand of $M_i(r_j)$ for all r_j, but since $M_i(r_3)$ is a lens space while $M_i(r_1)$ does not have a lens space summand, this is impossible.

By Lemma 22.2 $M_i(r_3)$ is a lens space, so if M_i is Seifert fibered then the orbifold of M_i is a disk with two cone points, hence $M_i(r_1)$ is either a connected sum of two lens spaces or a Seifert fibered space with orbifold a sphere with at most three cone points. This is impossible because by Lemma 22.2 $M_i(r_1)$ is of type $X(a_1, b_1; a_2, b_2)$ with some a_i or b_i greater than 2, which is irreducible and contains a separating essential torus, at least one side of which is not an I-bundle.

If M_i contains a non-separating torus then the lens space $M_i(r_3)$ would contain a non-separating surface, which is absurd.

(2) If M_i is non-hyperbolic then by (1) it has a non-trivial JSJ decomposition. By [MeS] we may choose the JSJ decomposition surfaces F to be α-equivariant. If we define a graph G with the components of $M_i|F$ as vertices and the components of F as edges connecting adjacent components of $M_i|F$, then the fact that M_i contains no non-separating torus implies that G is a tree. Let T be a component of F corresponding to an arc incident to a vertex v of valence 1 in G. Then T bounds the manifold W corresponding to v, which by definition of the JSJ decomposition is either Seifert fibered or hyperbolic. □

LEMMA 23.4. *Suppose M_i is non-hyperbolic and let T be the essential torus in M_i given in Lemma 23.3(2). Let X, W be the components of $M_i|T$, where $X \supset T_0$. If T is compressible in $M_i(r_a)$ for some $a = 1, 2$, then both $X(r_a)$ and $X(r_3)$ are solid tori, and W is hyperbolic.*

PROOF. T is compressible in $X(r_3)$ because $M_i(r_3)$ is a lens space. By assumption T is also compressible in $M_i(r_a)$. Since $M_i(r_a)$ contains no lens space summand, by Lemma 23.2 either both $X(r_a)$ and $X(r_3)$ are solid tori, or $X(r_a)$ is a solid torus and $X(r_3) = (S^1 \times D^2) \# L(p, q)$ for some $p > 1$. We need to show that the second case is impossible.

Let m_j be the slope on T which bounds a disk in $X(r_j)$, $j = a, 3$. Then $M_i(r_3) = W(m_3) \# L(p, q)$. Since $M_i(r_3)$ is a lens space $L(p, q)$, we have $W(m_3) = S^3$, so W is the exterior of a knot in S^3. If W is Seifert fibered then it is the exterior of a torus knot, so $M_i(r_a) = W(m_a)$ is obtained by Dehn surgery on a torus knot in S^3 and hence contains no separating essential torus, contradicting

Lemma 23.1. Since by definition W is Seifert fibered or hyperbolic, this implies that W is hyperbolic. Note that by Lemma 22.2 $p \geq 3$ for all $i \in \{6, ..., 13\}$. By Lemma 23.2(2) we have $\Delta(m_a, m_3) \geq p \geq 3$. Since $W(m_a) = M_i(r_a)$ is toroidal and $W(m_3) = S^3$, this is a contradiction to [GLu1, Theorem 1.1], which shows that only integral or half integral surgeries on hyperbolic knots in S^3 can produce toroidal manifolds. This completes the proof that both $X(r_a)$ and $X(r_3)$ are solid tori.

If W is not hyperbolic then by definition W is Seifert fibered. By the above $X(r_3)$ is a solid torus. Let m_3 be a meridian slope of $X(r_3)$. Then $M_i(r_3) = W(m_3)$, so $M_i(r_3)$ being a lens space implies that the orbifold of W is a disk with two cone points, in which case $M_i(r_a) = W(m_a)$ is either a connected sum of two lens spaces or a Seifert fiber space with orbifold a sphere with at most three cone points. In the first case $M_i(r_a)$ contains no essential torus, while in the second case the only possible essential torus in $M_i(r_a)$ is a horizontal torus cutting the manifold into a $T^2 \times I$, or two twisted I-bundles over the Klein bottle. This is a contradiction because by Lemma 23.1 $M_i(r_a)$ contains an essential torus cutting it into either a Seifert fiber space with orbifold an annulus with a cone point of index 2, or two Seifert fiber spaces, at least one of which is not a twisted I-bundle over the Klein bottle. □

LEMMA 23.5. *The torus T in Lemma 23.3(2) is incompressible in both $M_i(r_1)$ and $M_i(r_2)$.*

PROOF. First assume that T is compressible in both $X(r_1)$ and $X(r_2)$. By Lemma 23.4 $X(r_j)$ is a solid torus for $j = 1, 2, 3$. Since $\Delta(r_1, r_2) > 1$, by Lemma 23.2 we see that X is the exterior of a (p, q) knot in a solid torus. Since T is not boundary parallel, $p > 1$. Let r be the cabling slope on T_0. Since $X(r_j)$ is a solid torus, we have $r_j \neq r$. Therefore by [CGLS, Theorem 2.4.3] we must have $\Delta(r, r_j) = 1$ for $i = 1, 2, 3$. By Lemma 23.2 one can check that $\Delta(r_1, r_3) = 1$ and $\Delta(r_2, r_3) \leq 2$. Since $\Delta(r_1, r_2) \geq 4$, this is a contradiction because any three slopes r_1, r_2, r_3 with distance 1 from a given slope r have the property that $\Delta(r_a, r_b) + \Delta(r_b, r_c) = \Delta(r_a, r_c)$ for some permutation (a, b, c) of $(1, 2, 3)$.

Now assume that T is compressible in $M_i(r_1)$, say. By Lemma 23.4 W is hyperbolic. On the other hand, by the above T is incompressible in $M_i(r_2)$, so W is a submanifold in $M_i(r_2)$ bounded by an incompressible torus, hence by Lemma 23.1(3) it is non-hyperbolic, which is a contradiction. □

LEMMA 23.6. *Let $T(a_1, b_1; a_2, b_2) = (S^3, L)$, where $a_i, b_i \geq 2$. If at least one of a_1, b_1, a_2, b_2 is greater than 2 then the exterior of L is atoroidal, and there is no Möbius band F in S^3 with $F \cap L$ a component of L.*

PROOF. Denote by $T(a)$ a rational tangle with slope $1/a$, where a is an integer. Given a tangle $\tau = (B^3, \tau)$, denote $B^3 - \text{Int} N(\tau)$ by E or $E(\tau)$, and call it the tangle space of τ. Since $T(a)$ is a trivial tangle in the sense that τ is rel ∂ isotopic to arcs on ∂B^3, the tangle space $E(a)$ is atoroidal, and any incompressible annulus in $B^3 - \tau$ is trivial in the sense that it is either parallel to an annulus on $\partial(B^3 - \tau)$ or cuts off a $D^2 \times I$ in B^3 with $\tau \cap (D^2 \times I)$ a core arc.

The tangle space $E(r_1, r_2)$ of a Montesinos tangle $T(r_1, r_2)$ is obtained by gluing $E(r_1), E(r_2)$ along a twice punctured disk $P = E(r_1) \cap E(r_2)$. The above implies that $E(r_1, r_2)$ is always atoroidal. If A is an essential annulus in $E(r_1, r_2)$ with minimal intersection with P, then an innermost circle outermost arc argument shows that A intersects P in essential arcs or circles in A. If the intersection is a set of circles then each component of $A \cap E(r_i)$ is a set of trivial annuli, which implies that A is also trivial. If each component of $A \cap P$ is an essential arc then each component of $A \cap E(r_i)$ is a bigon in the sense that it is a disk intersecting P in two arcs, which implies that $r_i = 2$ for $i = 1, 2$. Therefore $E(r_1, r_2)$ contains no essential annulus unless $r_1 = r_2 = 1/2$ mod 1.

By definition $T(a_1, b_1; a_2, b_2)$ is the union of two Montesinos tangles $T(a_i, b_i)$. If the tangle space of $T(a_1, b_1; a_2, b_2)$ is toroidal then either one of the $T(a_i, b_i)$ is toroidal or they are both annular. By the above neither case is possible if at least one of a_1, b_1, a_2, b_2 is greater than 2.

The proof for a Möbius band is similar. If F is a Möbius band in S^3 bounded by a component of L and has interior disjoint from L then after cutting along the surface $P_1 = E(a_1, b_1) \cap E(a_2, b_2)$ it either lies in one of the $E(a_i, b_i)$ or intersects each in bigons. One can show that the first case is impossible, and in the second case $a_i = b_i = 2$ for $i = 1, 2$. □

LEMMA 23.7. *M_i is hyperbolic for $i = 6$ or $8 \leq i \leq 13$.*

PROOF. Let T be the α-equivariant essential torus in M_i given in Lemma 23.3(2). By Lemma 23.5 T is incompressible in both $M_i(r_a)$, $a = 1, 2$. Since T is α-equivariant, its image F in $Q_i = M_i/\alpha$ is a 2-dimensional orbifold with zero orbifold Euler characteristic (see [Sct] for definition), and all the cone points have indices 2. Hence it is T^2, K^2, $P^2(2,2)$, $S^2(2,2,2,2)$, A^2, M^2, or $D^2(2,2)$, where the surfaces are torus, Klein bottle, projective plane, sphere, annulus, Möbius band and disk, and the numbers indicate the indices of the cone points. Note that in the last three cases the boundary of the surface is part of the branch set of α. Since T is incompressible in $M_i(r_a)$, F is incompressible in $Q_i(r_a)$ in the sense that if some simple loop on F bounds a disk in $Q_i(r_a)$ intersecting the branch set at most once then it bounds such a disk on F. We need to show that for each type of surface above there is some $a = 1, 2$ such that no such incompressible 2-dimensional orbifold exists in $Q_i(r_a)$.

We have $Q_i = (B^3, K_i)$, where $B^3 = M_i/\alpha$ is a 3-ball and K_i is the branch set of α. Since F lies in B^3, it cannot be K^2 or P^2. For all i one can check that the branch set K_i of α in Q_i contains at most one closed circle, hence the case A^2 is also impossible.

By Lemma 22.2, $Q_i(r_1) = T(a_1, b_1; a_2, b_2)$ for some $a_j, b_j \geq 2$, and $(a_j, b_j) \neq (2,2)$ for some j. Therefore by Lemma 23.6 we have $F \neq T^2, M^2$ for $i = 6, ..., 13$ because there are no such surfaces in $Q_i(r_1)$. It remains to show that $F \neq D^2(2,2)$ or $S^2(2,2,2,2)$.

For $i = 13$, $Q_i(r_2) = (S^3, L)$, where L consists of two parallel copies of a trefoil knot K. Since each component of L is non-trivial in S^3, $F \neq D^2(2,2)$ in this case. Suppose $F = S^2(2,2,2,2)$, and let V be a regular neighborhood of the trefoil knot containing L, intersecting F minimally. Then $F \cap V \neq \emptyset$, and F is not contained in V as otherwise one can show that $F - L$ would be compressible. Therefore $F \cap V$ is a union of two meridian disks, and $F \cap S^3 - \text{Int} V$ is an essential annulus in $S^3 - \text{Int} V$. Since $S^3 - \text{Int} V$ contains no essential annulus with the meridian of V as boundary slope, this is a contradiction.

The proofs for the cases $i \in \{6, 8, 9, 10, 11, 12\}$ are similar. In these cases $Q_i(r_2) = (S^3, L)$, and there is a torus T' cutting S^3 into two solid tori V_1, V_2, each containing some components of L. One can check that no component L' of L bounds a disk intersecting $L - L'$ at two points, so $F \neq D^2(2,2)$. If $F = S^2(2,2,2,2)$ then either F lies in one of the V_j, or it intersects one of the V_j in two meridional disks and the other V_k in an essential annulus with boundary slope the meridional slope of V_j. Neither case is possible for the $Q_i(r_2)$ listed in Lemma 22.2. □

LEMMA 23.8. *M_7 is hyperbolic.*

PROOF. By Lemma 22.2 we have $M_7(r_1) \in X(3,3;2,3)$ and $M_7(r_2) \in X(2,2;2,3)$. Consider the tangle decomposition sphere P_a of the orbifold $Q_7(r_a)$, $a = 1, 2$, which corresponds to a horizontal plane in Figure 22.7(b), (d) respectively. It lifts to an essential torus T_a in $M_7(r_a)$.

Each side of P_a is a Montesinos tangle of type $T(r_1, r_2)$, which is the sum of two rational tangles over a disk D. The boundary of D determines the fibration of the double branched cover $X(r_1, r_2)$ of $T(r_1, r_2)$, which has a unique Seifert fibration unless $r_1 = r_2 = 2$, in which case the closed circle in the tangle is isotopic (without crossing the arcs) to a curve on the punctured sphere, which lifts to a fiber in the other fibration of $X(r_1, r_2)$. It is easy to check from Figures 22.7(b) and (d) that the fiber curves from the two sides of P_a do not match, so $M_7(r_a)$ is not a Seifert fiber space. Since each side of T_a is a small Seifert fiber space with orbifold a disk with two cone points, it follows that $M_7(r_a)$ contains no other essential torus.

Suppose M_7 is non-hyperbolic and let T be the essential torus in M_7 given by Lemma 23.3. By Lemma 23.5 it is incompressible in both $M_7(r_a)$, therefore by the uniqueness of T_a above we see that $T = T_a$ in $M_7(r_a)$ up to isotopy. As before, denote by W and X the components of $M_7|T$, with $X \supset T_0$. Then W is the manifold on one side of T_a in $M_7(r_a)$. Therefore we must have $W = X(2,3)$, so $X(r_1) = X(3,3)$ and $X(r_2) = X(2,2)$. We will show that this is impossible.

Let Y be the component of the JSJ decomposition of X that contains T. Then Y is either hyperbolic or Seifert fibered. There are three cases.

Case 1. *$T_0 \subset \partial Y$ and Y is Seifert fibered.* By Lemma 23.5 T is incompressible in $Y(r_a)$ for $a = 1, 2$, so r_a is not the fiber slope on T_0. Hence the Seifert fibration extends over $Y(r_1)$ and $Y(r_2)$. In this case $\partial Y - T_0$ is incompressible in $Y(r_a)$. Since $X(r_a)$ is atoroidal, either $Y(r_1) \cong Y(r_2) \cong T^2 \times I$, or $Y = X$. In the first case we have $X(r_1) = X(r_2)$, which is a contradiction because $X(r_1) = X(3,3) \not\cong X(2,2) = X(r_2)$. In the second case, Since $X(r_1)$ has orbifold $D^2(3,3)$, the orbifold of X must be $A^2(3,3)$ or $A^2(3)$. On the other hand, since $X(r_2)$ has orbifold $D^2(2,2)$, the orbifold of X must be $A^2(2,2)$ or $A^2(2)$, which contradicts the fact that Seifert fibrations for these manifolds are unique.

Case 2. *$T_0 \subset \partial Y$ and Y is hyperbolic.* If ∂Y has more than two boundary components then the fact that $X(r_a)$ is atoroidal implies that $Y(r_a)$ is either ∂-reducible, or a $T^2 \times I$. If $\partial Y = T \cup T_0$ then $Y = X$ and by assumption both $Y(r_a) = X(r_a)$ are annular and atoroidal. In either case $Y(r_a)$ is either ∂-reducible or annular and atoroidal. Since $\Delta(r_1, r_2) = 5$ and Y is hyperbolic, this is a contradiction to [Wu2, Theorem 1.1] if both $Y(r_a)$ are ∂-reducible, to [GW2] if one of them is ∂-reducible and the other is annular, and to [GW3] if both $Y(r_a)$ are annular and atoroidal. (The main theorem of [GW3] said that if $\Delta = 5$ then Y and r_1, r_2 are listed in one of the three possibilities in [GW3, Theorem 1.1], but in that case both $Y(r_a)$ are toroidal.)

Case 3. *$T_0 \not\subset \partial Y$.* Let X_1 be the component of $X|(\partial Y)$ containing T_0, and let X_2 be the closure of $X - X_1$, so $X = X_1 \cup X_2$. Since M_7 contains no non-separating torus (Lemma 23.3), $T_1 = X_1 \cap X_2$ is a single torus. Since $X(r_a)$ is atoroidal, T_1 must be compressible in $X_1(r_a)$ for $a = 1, 2$. Thus Lemma 23.2 and the fact that $X(r_a)$ contains no lens space summand imply that $X_1(r_a)$ is a solid torus for $a = 1, 2$, as in Lemma 23.2(1); moreover, since $\Delta(r_1, r_2) > 1$, by Lemma 23.2(1) X_1 is a (p, q) cable space, and by [CGLS, Theorem 2.4.3] $\Delta(r, r_a) = 1$ for r the cabling slope. It is easy to see that the meridian slopes m_a of $X_1(r_a)$ satisfy $\Delta(m_1, m_2) = p\Delta(r_1, r_2) \geq 5$. Now $X(r_a) = X_2(m_a)$, so by Cases 1 and 2 above applied to X_2 we see that this case is also impossible. □

Denote by $c \cdot d$ the minimal intersection number between the two isotopy classes of simple closed curves on a surface represented by c and d, respectively. If $\varphi : F \to F$ is a homeomorphism and K a curve on the surface F, then K is said to be φ-full if for any essential curve c on F there is some i such that $c \cdot \varphi^i(K) \neq 0$.

If K is a knot in a 3-manifold Y with a preferred meridian-longitude, denote by $Y(K, p/q)$ the manifold obtained from Y by p/q surgery on K. Let $X = F \times I/\psi$ be an F-bundle over S^1 with gluing map ψ, let $F_t = F \times t$, $t \in I = [0, 1]$, and let K be an essential curve on $F_{1/2}$. Then there is a preferred meridian-longitude pair (m, l) on $\partial N(K)$, with l the slope of $F_{1/2} \cap \partial N(K)$.

LEMMA 23.9. *Let $X = F \times I/\psi$. Let $\eta : F \times I \to F_0$ be the projection, $\varphi = \eta \circ \psi$, and K an essential curve on $F_{1/2}$. If $\eta(K) \subset F_0$ is φ-full and $q > 1$, then $X(K, p/q)$ is hyperbolic.*

PROOF. Let A_i be an annulus in X with $\partial A_i = K \cup K_i$, where $i = 0, 1$ and $K_i \subset F_i$. Let V_i be a regular neighborhood of A_i. Put $Y = F \times I$. Then $Y = V_1 \cup W$, where W is homeomorphic to $F \times I$, and $V_1 \cap W$ is an annulus A'. After p/q surgery on K we have $Y(K, p/q) = V_1(K, p/q) \cup W$. Note that $V_1(K, p/q)$ is a solid torus with A' an annulus on $\partial V_1(K, p/q)$ running q times along the longitude. By an innermost circle outermost arc argument one can show that $Y(K, p/q)$ is irreducible, ∂-irreducible, atoroidal, and any essential annulus A_2 can be isotoped to be disjoint from K_1, i.e. $\partial A_2 \cdot K_1 = 0$. Moreover, if A_2 has at least one boundary component on F_1 then A_2 is either vertical in the sense that it is isotopic to $c \times I \subset F \times I$ for some curve $c \subset F$, or isotopic to A' and hence has both boundary curves parallel to K_1. Similarly, using A_0 and V_0 one can show that $\partial A_2 \cdot K_0 = 0$, and if A_2 is not vertical and $A_2 \cap F_0 \neq \emptyset$ then it has both boundary curves parallel to K_0.

The above facts imply that $X(K, p/q) = Y(K, p/q)/\psi$ is irreducible. Since the non-separating surface F_0 cuts $X(K, p/q)$ into $Y(K, p/q)$, which is not an I-bundle, we see that $X(K, p/q)$ is not Seifert fibered. It remains to show that $X(K, p/q)$ is atoroidal.

If T is an essential torus in $X(K, p/q)$ then it can be isotoped so that $T \cap Y(K, p/q) = Q$ is a set of essential annuli. Let $C_i = Q \cap F_i$. We claim that for any curve $c \subset C_0$, $\varphi(c)$ is isotopic to a curve in C_0.

We have $\psi(C_0) = C_1$, so $\psi(c) \subset C_1$. If $\psi(c)$ belongs to a vertical annulus Q_1 then $\varphi(c) = \eta(\psi(c)) \cong Q_1 \cap F_0 \subset C_0$. If $\psi(c)$ belongs to a non-vertical annulus then by the property proved above, $\psi(c)$ is isotopic to K_1, so $\varphi(c) \cong \eta(K_1) = K_0$. Note that if Q has a non-vertical component with boundary on F_1 then the fact that C_0, C_1 have the same number of components implies that there is also a non-vertical component Q_0 with boundary on F_0, and we have shown that each component c' of ∂Q_0 is isotopic to K_0, so $\varphi(c) \cong c' \subset \partial Q_0 \subset C_0$. This completes the proof of the claim.

Let c be a component of C_0. We have shown above that $c \cdot K_0 = 0$ for any $c \subset C_0$. Applying the above to φ^{-1} we see that there is a curve $c' \subset C_0$ such that $\varphi(c') \cong c$. By induction we have $c \cdot \varphi^i(K_0) = c' \cdot \varphi^{i-1}(K_0) = 0$ for all i, which is a contradiction to the assumption that $K_0 \cong \eta(K)$ is φ-full and hence $c \cdot \varphi^i(K_0) \neq 0$ for some i. □

LEMMA 23.10. *The manifold M_5 is hyperbolic.*

PROOF. Let $W = M_5|F_b$, let F_+, F_- be the two copies of F_b in W, and let A be the annulus $T_0|\partial F_b$. Then W is obtained from $Y = F_+ \cup F_- \cup A$ by attaching faces of Γ_a and then some 3-cells.

Two faces of Γ_a are *parallel* if their boundary curves are parallel on Y. Since parallel faces cobound a 3-cell in W, we need only attach one such face among a set of parallel faces. From Figure 11.10 one can check that the four bigons are parallel faces, and the two 6-gons are parallel to each other. Therefore W is obtained from Y by attaching one bigon σ_1, one 6-gon σ_2 and then a 3-cell. Let σ_1 be the bigon on Figure 11.10 between the edges B and G, and assume that the edge $B \subset F_+$.

Cutting W along σ_1, we obtain a manifold W_1 with boundary a torus, and it contains the 6-gon σ_2. Therefore it is a solid torus such that the remnant of F_+, denoted by F'_+, runs along the longitude three times. If we replace σ_2 and the attached 3-cell by a solid torus J with meridian intersecting F'_+ in one essential arc then W_1 becomes a $F'_+ \times I$ and W becomes $X = F_+ \times I$. Therefore $W = X(K, p/q)$, where K is the core of J, and $q = 3$. Let $\psi : F_- \to F_+$ be the gluing map, $\eta : F_+ \times I \to F_-$ the projection, and $\varphi = \eta \circ \psi$. By Lemma 23.9 we need only show that the curve K_- on F_- isotopic to K is φ-full.

In M_5 the bigon σ_1 has boundary edges $B \cup G$ on F_b, as shown in Figure 11.10(b). Suppose $B \subset F_+$ and $G \subset F_-$ when we consider σ_1 as a bigon in $F_+ \times I$. Then ψ maps the curve B on F_- to the curve B on F_+, which is mapped to G on F_- by η. Therefore $\varphi : F_- \to F_-$ maps B to G. Since $F_-|B$ is an annulus and B is disjoint from the curve K_- above, this determines K_-. Also $\varphi(K_-)$ is the curve on F_- disjoint from $\varphi(B) = G$, so K_- intersects $\varphi(K_-)$ transversely at a single point, cutting F_- into an annulus. Therefore K_- is φ-full, and the result follows. □

LEMMA 23.11. *The manifold M_4 is hyperbolic.*

PROOF. The proof is similar to that of Lemma 23.10. In this case $W = M_4|F_b$ is obtained from $F_+ \cup F_- \cup A_1 \cup A_2$ by attaching two bigons σ_1, σ_2 and one 4-gon σ_3, so $W = X(K, p/q)$ with $q = 2$, where $X = F_+ \times I$ and K is disjoint from σ_1, σ_2. Choose σ_1, σ_2 to be the bigons in Figure 11.9(a) bounded by $H \cup E$ and $E \cup N$, respectively. Then F_- can be identified with F_b, and σ_1, σ_2 intersects F_- in the edges E and N, respectively. These cut F_- into an annulus containing the curve K_- isotopic to the knot K in $X = F_+ \times I$. The map $\varphi : F_- \to F_-$ maps the edges E and N in Figure 11.9(b) to H and E, respectively, so $\varphi(K_-)$ is the curve in the annulus $F_-|(H \cup E)$. The curves K_- and $\varphi(K_-)$ intersect transversely at a single point, cutting F_- into a neighborhood of ∂F_-, hence K_- is φ-full, and M_4 is hyperbolic by Lemma 23.9. □

LEMMA 23.12. *Let F be a closed orientable surface of genus 2, and let α, β be two non-separating simple closed curves on F, intersecting minimally, cutting F into disks. Let X be obtained from $F \times I$ ($I = [0,1]$) by attaching a 2-handle along $\alpha \times \{0\}$. Identify F with $F \times \{1\} \subset F \times I$, and let $T = \partial X - F$. Then*

(1) a compressing disk D of F intersects β at least 3 times; and

(2) an incompressible annulus A in X with $\partial A \subset F$ and $\partial A \cap \beta = \emptyset$ is boundary parallel.

PROOF. (1) Let E be the disk in X bounded by $\alpha \times 1$, cutting X into $X' = T \times I$. Note that X is a compression body, and $\{E\}$ is the unique (up to isotopy) complete disk system for X.

By assumption $\alpha \cup \beta$ cuts F into disks, hence $|\alpha \cap \beta| \geq 3$. Since α intersects β minimally, we may choose a hyperbolic structure on the surface F so that α, β are geodesics. Let D be a compressing disk for F in X. Up to isotopy we may assume that $\gamma = \partial D$ is a geodesic or a slight push off of a geodesic if it is isotopic to α or β. Then both $|\gamma \cap \alpha|$ and $|\gamma \cap \beta|$ are minimal up to isotopy.

We may assume that $D \cap E$ consists of arcs. If $D \cap E \neq \emptyset$, by taking an arc that is outermost on D, surgering E along the corresponding outermost disk in D, and discarding one of the resulting components, we get a new disk E' having fewer intersections with D, such that $\{E'\}$ is a complete disk system for X. Since $\{E\}$ is the unique complete disk system for X, E' is isotopic to E. Since $|\partial E' \cap \partial D| < |\partial E \cap \partial D| = |\alpha \cap \gamma|$ and $\partial E'$ is isotopic to ∂E, this is a contradiction to the fact that $|\gamma \cap \alpha|$ is minimal up to isotopy. Therefore $D \cap E = \emptyset$. Hence D either (a) is parallel to E, or (b) cuts off a solid torus containing E.

In case (a) ∂D is a parallel copy of α, so $|\partial D \cap \beta| = |\alpha \cap \beta| \geq 3$ and we are done. In case (b), let F_1 be the punctured torus on F bounded by γ which does not contain α. If $|\gamma \cap \beta| \leq 2$ then F_1 contains at most one arc of β, so it contains an essential loop disjoint from $\alpha \cup \beta$, which is a contradiction to the assumption that $\alpha \cup \beta$ cuts F into disks.

(2) Let A be an incompressible annulus in X with $\partial A \subset F - \beta$. We may assume that α, β are hyperbolic geodesics, and each component of ∂A is either a geodesic, or a slight push off of a geodesic if it is parallel to α, β or another component of ∂A. Thus both $|\partial A \cap \alpha|$ and $|\partial A \cap \beta|$ are minimal; in particular, $\partial A \cap \beta = \emptyset$. As in (1), this implies that $A \cap E$ consists of essential arcs on A. If $A \cap E = \emptyset$ then ∂A lies in $F|(\alpha \cup \beta)$, but since A is incompressible while each component of $F|(\alpha \cup \beta)$ is a disk, this is impossible. Therefore we may assume that $A \cap E$ is a non-empty set C of essential arcs on A.

Let B_i be a component of $A|C$. Then B_i is a disk in $X' = T \times I$, so ∂B_i is a trivial loop on T', bounding a disk B'_i on T'. Let E_1, E_2 be the two copies of E on T'. If $B'_i \cap (E_1 \cup E_2)$ is a single disk then one can use a disk component of $B'_i \cap F$ to isotope A to reduce $|\partial A \cap \partial E| = |\partial A \cap \alpha|$, which is a contradiction to the minimality of $|\partial A \cap \alpha|$. Therefore $B'_i \cap (E_1 \cup E_2)$ consists of two disks, and $B' \cap F$ is a single disk P_i. One can check that $\cup P_i$ is an annulus on F parallel to A. □

LEMMA 23.13. *The manifold M_{14} is hyperbolic.*

PROOF. Cutting M_{14} along the surface F_b, we obtain two manifolds X_1, X_2, where X_1 is the one containing the four bigon faces of F_a, and X_2 contains the two 4-gon faces of F_a. Let σ_1, σ_2 be the bigons on F_a bounded by the edges $E \cup F$ and $B \cup Y$ respectively in Figure 20.6, and let σ_3 be the 4-gon bounded by the edges $B \cup C \cup Y \cup F$. Note that any other face of F_a is parallel in X_i to one of these.

Let $A_i = X_i \cap T_0$. Then X_1 is obtained from the genus 2 surface $F_b \cup A_1$ by attaching σ_1, σ_2 and then a 3-cell, hence it is a handlebody of genus 2 because $\partial \sigma_1, \partial \sigma_2$ are disjoint nonparallel nonseparating curves on $F_b \cup A_1$. The core of A_1

is a curve on ∂X_1 such that after attaching a 2-handle to X_1 along A_1 we get the manifold on the side of $\hat{F}_b$ which contains no torus boundary component, hence from Figure 22.14(b) we see that it is the double branched cover of a Montesinos tangle $T(2,2)$, which is a twisted I-bundle over the Klein bottle. This implies that the surface $F_b = \partial X_1 - A_1$ is incompressible in X_1.

Now consider X_2. Let F be the genus 2 surface $F_b \cup A_2$, α the boundary of σ_3, and β the core of A_2. Then α intersects β minimally at four points. From Figure 20.6(b) we see that the edges B, C, Y, F cut the surface F_b into two disks, hence $\alpha \cup \beta$ cuts F into disks. X_2 is obtained from $F \times I$ by attaching a 2-handle along the curve $\alpha \times \{0\}$. Therefore it satisfies the conditions of Lemma 23.12. In particular, F_b is incompressible in X_2.

Since X_1 is a handlebody and X_2 is a compression body, they are irreducible and atoroidal. Since M_{14} is obtained by gluing X_1, X_2 along the incompressible surface F_b, M_{14} is also irreducible. It is well known that an incompressible surface in a Seifert fiber space is either vertical, and therefore an annulus or torus, or horizontal, in which case it intersects all boundary components. Since the surface F_b satisfies neither condition, we see that M_{14} is not Seifert fibered. It remains to show that M_{14} is atoroidal.

Assume M_{14} is toroidal and let T_1 be an essential torus in M_{14} intersecting F_b minimally. Since X_i is atoroidal, T_1 intersects X_i in incompressible annuli. A component A'_2 of $T_1 \cap X_2$ is an incompressible annulus in X_2 disjoint from β, hence by Lemma 23.12 it is parallel to an annulus A'' on ∂X_2. If $A'' \subset F_b$ then T_1 can be isotoped to reduce $|T_1 \cap F_b|$, which is a contradiction to the minimality assumption. Therefore $A'' \supset \beta$ and hence $A'' \supset A_2$, so each component of $\partial A''$ is parallel to a component of ∂F_b. Since this is true for all components of $T_1 \cap X_2$, we see that each component of $T_1 \cap F_b$ is parallel to a component of ∂F_b.

Now let A'_1 be a component of $T_1 \cap X_1$. By the above, the two boundary components of A'_1 are parallel on ∂X_1. Since X_1 is a handlebody, A'_1 is parallel to an annulus A''_2 on ∂X_1. For the same reason as above, it must contain the annulus A_1. This is true for all components of $T_1 \cap X_1$. Let A'_i be a component of $T_1 \cap X_i$ which is closest to A_i. Then $\partial A'_1 = \partial A'_2$, hence $T_1 = A'_1 \cup A'_2$. It follows that T_1 is parallel to T_0, contradicting the assumption that T_1 is essential in M_{14}. □

THEOREM 23.14. *The manifolds M_i in Definition 21.3 are all hyperbolic.*

PROOF. This follows from [GW1, Theorem 1.1] for $i = 1, 2, 3$, and from Lemmas 23.7, 23.8, 23.10, 23.11 and 23.13 for $i > 3$. □

24. Toroidal surgery on knots in S^3

Recall that each of the manifolds M_1, M_2, M_3 admits two toroidal Dehn fillings r'_i, r''_i on a torus boundary component T_0 with distance 4 or 5. These are the exteriors of the links L_1, L_2, L_3 in Figure 24.1. Let $L_i = K'_i \cup K''_i$, where K'_i is the left component of L_i. Let $T_1 = \partial N(K'_i)$, and $T_0 = \partial N(K''_i)$.

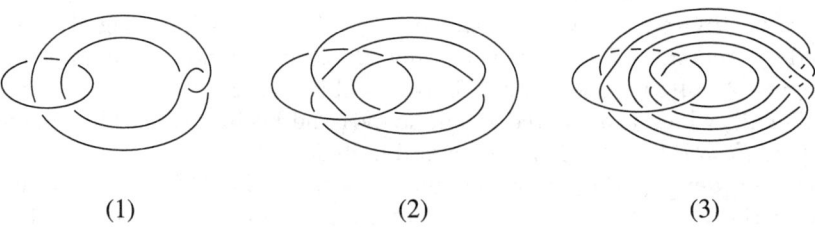

(1) (2) (3)

Figure 24.1

Each M_i has a pair of toroidal slopes r'_i, r''_i on T_0. These are given in [GW1, Theorem 7.5] and shown in Figures 7.2, 7.4 and 7.5 of [GW1].

LEMMA 24.1. *With respect to the preferred meridian-longitude pair of K''_i, the slopes r'_i, r''_i are given as follows, up to relabeling.*
 (1) $r'_1 = 0$ and $r''_1 = 4$.
 (2) $r'_2 = -2$ and $r''_2 = 2$.
 (3) $r'_3 = -9$ and $r''_3 = -13/2$.

PROOF. (1) This is basically proved in [GW1, Lemma 7.1]. It was shown that M_1 is the double branched cover of the tangle Q_1 in Figure 7.2(c) of [GW1]. Let m be the meridian, l the preferred longitude, and l' the blackboard longitude of the diagram of K''_1 in [GW1, Figure 7.2(a)]. Calculating the linking number of l' with K''_1 in Figure 7.2(a) we see that $l' = 2m + l$. Let $\eta : M_1 \to Q_1$ be the branched covering map. If r is a slope on T_0 then $\eta(r)$ is a curve of a certain slope on the inside boundary sphere, which will be denoted by a number in $\mathbb{Q}_0 = \mathbb{Q} \cup \{\infty\}$. One can check that $\eta(m) = 0/1$, and $\eta(l') = 1/0$. The two toroidal slopes r'_1, r''_1 map to slopes $-1/2$ and $1/2$, as shown in Figure 7.2(d) and (e) of [GW1]. We have $\varphi(-2m+l') = (-2 \times 0 + 1)/(-2 \times 1 + 0) = -1/2$, and $\varphi(2m + l') = (2 \times 0 + 1)/(2 \times 1 + 0) = 1/2$. Therefore $r'_1 = -2m + l' = -2m + (2m + l) = l$ and $r''_1 = 2m + l' = 4m + l$.

(2) This is similar to (1), using [GW1, Figure 7.4] instead. We have $\eta(m) = 0/1$, $\eta(l') = 1/2$, $\eta(r'_2) = 1/0$, $\eta(r''_2) = 1/4$, and $l = l'$. Therefore $r'_2 = -2m + l$ and $r''_2 = 2m + l$.

(3) Use 7.5(k) to denote [GW1, Figure 7.5(k)]. 7.5(a) shows that $l' = l - 6m$. A careful tracking of l' during the modification from 7.5(a) to 7.5(b) then to 7.5(c) shows that $\eta(m) = 0/1$ and $\eta(l') = 1/3$. From 7.5(c) and 7.5(e) we see that $\eta(r'_3) = 1/0$ and $\eta(r''_3) = 2/5$. Therefore $r'_3 = -3m + l' = -9m + l$, and $r''_3 = -m + 2l' = -m + 2(l - 6m) = -13m + 2l$. □

LEMMA 24.2. *Suppose K is a hyperbolic knot in S^3 admitting two toroidal Dehn surgeries $K(r_1), K(r_2)$ with $\Delta(r_1, r_2) = 4$ or 5. Then there is an $i \in \{1, 2, 3\}$ and a slope s on T_1 of M_i such that $(E(K), r_1, r_2) \cong (M_i(s), r'_i, r''_i)$.*

PROOF. Let F_a be an essential punctured torus in $M_K = S^3 - \text{Int} N(K)$ such that $\hat{F}_a$ is an essential torus in $K(r_a)$, chosen so that $|\partial F_a|$ is minimal. By Theorem 21.4 the triple $(E(K), r_1, r_2)$ is equivalent to either (M_i, r'_i, r''_i) with $1 \leq i \leq 14$, or to $(M_i(s), r'_i, r''_i)$ for some $i = 1, 2, 3, 14$. Therefore we need only show that the manifold M_i ($i = 4, ..., 14$) is not the exterior of a knot or link in S^3.

When $i = 4$, the surface F_b has two boundary circles on T_0 with the same orientation. Let A be an annulus on T_0 connecting these two boundary components. Then $F_b \cup A$ is a non-orientable closed surface in M_4. It follows that M_4 cannot be the exterior of a knot in S^3 because S^3 contains no embedded non-orientable surface.

For $i = 5$, let V_b be the Dehn filling solid torus of $M_5(r_b)$. Then the $\mathbb{Z}_2$ homology group H of $V_b \cup F_b$ is generated by α, x and y, where α is the core of V_b, and x, y are represented by the edges A and E in Figure 11.10(b), respectively. A bigon in Figure 11.10(a) gives the relation $x = y$. Consider the quotient group H' obtained from H by identifying x with y. Then $H' = \mathbb{Z}_2 \oplus \mathbb{Z}_2$ is generated by α and x. Each corner of Figure 11.10(a) represents the element α, and each edge represents x in H'. Since each face in Figure 11.10(a) has an even number of edges and an even number of corners on its boundary, it represents 0 in H'. Therefore
$$H_1(M_5(r_b), \mathbb{Z}_2) = H_1(V_b \cup F_b \cup F_a, \mathbb{Z}_2) = H' = \mathbb{Z}_2 \oplus \mathbb{Z}_2.$$
Since the $\mathbb{Z}_2$ homology of any manifold obtained by Dehn surgery on a knot in S^3 is either trivial or $\mathbb{Z}_2$, it follows that M_5 is not a knot exterior.

Now assume that M_i is the exterior of a knot K in S^3 for some $i \geq 6$. By Theorem 23.14 K is hyperbolic. Put $\mathbb{Q}_0 = \mathbb{Q} \cup \{\infty\}$. A number in $\mathbb{Q}_0$ is represented by p/q, where p, q are coprime integers, and $q \geq 0$. Given a meridian-longitude pair (m, l) and $r = p/q \in \mathbb{Q}_0$, denote by $K(r)$ the manifold obtained by surgery on K along the slope $pm + ql$. There is a one to one correspondence $\eta : \mathbb{Q}_0 \to \mathbb{Q}_0$ such that $\Delta(\eta(r), \eta(s)) = \Delta(r, s)$, and $K(\eta(r))$ is the double branched cover of $Q_i(r)$, which is the manifold $X_i(r)$ given in Lemma 22.2. Since $K(\eta(r_3))$ is a lens space, by the Cyclic Surgery Theorem [CGLS, p.237] the slope $\eta(r_3)$ is an integer slope with respect to the preferred meridian-longitude of K. To simplify the calculation, let $l = \eta(r_3)$.

By [GLu1] $\eta(r_1)$ and $\eta(r_2)$ are integer or half integer slopes. Suppose $\eta(r_i) = p_i/q_i$. Then $p_3/q_3 = 0/1$. By the above we have $q_i = 1$ or 2 for $i = 1, 2$. By Lemma 22.2, $|p_1| = \Delta(r_1, r_3) = 1$, and $|p_2| = \Delta(r_2, r_3) \leq 2$.

If $|p_2| = 1$ then $4 \leq \Delta(r_1, r_2) = \Delta(\eta(r_1), \eta(r_2)) = |p_1 q_2 - p_2 q_1|$ implies that $q_1 = q_2 = 2$. This is a contradiction to [GWZ, Theorem 1], which says that a hyperbolic knot in S^3 admits at most one non-integral toroidal surgery.

We now have $|p_2| = 2$, so $\eta(r_2) = p_2/q_2 = \pm 2/1$. Since $\Delta(r_1, r_2) = |p_1 q_2 - p_2 q_1| = |\pm 1 - (\pm 2) q_1| \geq 4$, we must have $p_1/q_1 = \mp 1/2$, and $\Delta = 5$. From Lemma 22.2 we see that for $i \in \{6, ..., 13\}$, the only M_i satisfying $\Delta(r_2, r_3) = 2$ and $\Delta(r_1, r_2) = 5$ are the ones with $i = 7, 10$ or 11.

Consider the case $i = 10$. Let r_0 be the slope such that $\eta(r_0)$ is the meridian slope $1/0$. Then we have $\Delta(r_0, r_i) = \Delta(\eta(r_0), \eta(r_i)) = \Delta(1/0, p_i/q_i) = q_i$. Therefore by the above we have $\Delta(r_0, r_i) = q_i = 2, 1, 1$ for $i = 1, 2, 3$, respectively. By Lemma 22.2 we have $r_1 = 0/1$, $r_2 = -5/2$ and $r_3 = 1/0$. Let $r_0 = p'/q'$. Then we have
$$\Delta(r_0, r_1) = |p'| = 2$$
$$\Delta(r_0, r_2) = |2p' + 5q'| = 1$$
$$\Delta(r_0, r_3) = q' = 1$$
These equations have a unique solution $r_0 = -2/1$. One can check that $Q_{10}(-2/1)$ is the 2-bridge knot $K_{2/7}$, so its double branched cover is $L(7, 2) \neq S^3$, which is a contradiction.

The tangles Q_7, Q_{11} and Q_{14} in Figures 22.7, 22.11 and 22.14 have a circle component. If M_i is a knot exterior in S^3 then there is a slope r such that the double branched cover of $Q_i(r)$ is S^3. Since each of $Q_7(r)$ and $Q_{11}(r)$ has at least two components, its double branched cover has nontrivial $\mathbb{Z}_2$ homology [Sa, Sublemma 15.4], so M_7 and M_{11} are not knot exteriors in S^3. Similarly M_{14} is not the exterior of a link in S^3. □

LEMMA 24.3. *(1) Let $i \in \{1, 2, 3\}$. If r_1, r_2 are toroidal slopes of M_i on T_0 with $\Delta(r_1, r_2) \geq 4$, then $\{r_1, r_2\} = \{r_i', r_i''\}$.*

(2) The slope -7 is a solid torus filling slope on T_0 of M_3, and there is an orientation preserving homeomorphism of M_3 which interchanges the two solid torus filling slopes $\{1/0, -7/1\}$ and the two toroidal slopes $\{-9, -13/2\}$.

PROOF. Since M_1, M_2, M_3, M_{14} are the only ones in Definition 21.3 with two boundary components, by Theorem 21.4 (M_i, r_1, r_2) is equivalent to one of the (M_j, r_j', r_j'') with $j = 1, 2, 3, 14$. Since M_1, M_2, M_3 are link complements in S^3 and by Lemma 24.2 M_{14} is not, we have $j \neq 14$. Computing $H_1(M_j, T_1)$ shows $H_1(M_1, T_0) = \mathbb{Z}$, $H_1(M_2, T_0) = \mathbb{Z}_3$, and $H_1(M_1, T_0) = \mathbb{Z}_5$, hence we must have $j = i$.

By definition there is a homeomorphism $\varphi : (M_i, r_1, r_2) \to (M_i, r_i', r_i'')$, up to relabeling of r_1, r_2. For $i = 1, 2$, by [Ga1] and [Be] the knot K_i'' has no nontrivial solid torus surgery, hence $\varphi(m) = m$, where m is a meridian of K_i''. Since $\Delta(m, r_i') = \Delta(m, r_i'') = 1$, by the homeomorphism we also have $\Delta(m, r_1) = \Delta(m, r_2) = 1$, so r_1, r_2 are also integer slopes. It follows that if $\{r_1, r_2\} \neq \{r_1', r_1''\}$ then there is a pair of toroidal slopes with distance at least 5. Since M_1, M_2 is not homeomorphic to M_3, this is a contradiction to Theorem 21.4 and [Go].

Now suppose $i = 3$. By an isotopy one can deform the tangle in [GW1, Figure 7.5(c)], which is shown in Figure 24.2(a), to the one in Figure 24.2(b), which is invariant under the π rotation ψ along the forward slash diagonal. The $1/0$ slope in Figure 24.2(b) corresponds to the $1/2$ slope in Figure 24.2(a), which, by the proof of Lemma 24.1(3), lifts to the slope $-7m + l$ on T_0. The two toroidal slopes $1/0$ and $2/5$ for the tangle in Figure 24.2(a) correspond to the slopes $-1/2$ and 2 in Figure 24.2(b), which are interchanged by ψ. It follows that ψ lifts to an orientation preserving homeomorphism $\psi' : M_3 \to M_3$, which interchanges the two solid torus filling slopes $\{1/0, -7/1\}$ and toroidal slopes $\{-9, -13/2\}$. In fact, ψ' is represented by the matrix

$$A = \begin{pmatrix} 7 & 1 \\ -1 & 0 \end{pmatrix}$$

in the sense that if $A(p, q)^t = (p', q')^t$ (where B^t denotes the transpose of the matrix B) then $\psi'(pm + ql) = p'm + q'l$.

Solid torus surgeries on knots in a solid torus have been completely classified by Gabai [Ga1] and Berge [Be]. It was shown that there is only one knot admitting two nontrivial solid torus surgeries, which is a 7-braid. Since K_3'' is a 5-braid, we see that $m = 1/0$ and $m' = -7/1$ are the only solid torus filling slopes on T_0. Therefore the homeomorphism $\varphi : (M_3, r_3', r_3'') \to (M_3, r_1, r_2)$ must map the set of two curves $\{m, m'\}$ to itself, possibly with the orientation of one or both of the curves reversed. If φ preserves the orientation of m' and reverses the orientation of m then $\varphi(r_3') = 9/1$ would also be a toroidal slope, which is a contradiction to [Go] because $\Delta(-9/1, 9/1) = 18 > 8$. Similarly φ cannot preserve the orientation of m while reversing the orientation of m'. Therefore φ is orientation preserving and its induced map on the set of slopes on T_0 is either the identity map, which fixes $\{r_3', r_3''\}$, or the same as that induced by ψ' above, which interchanges $\{r_3', r_3''\}$. □

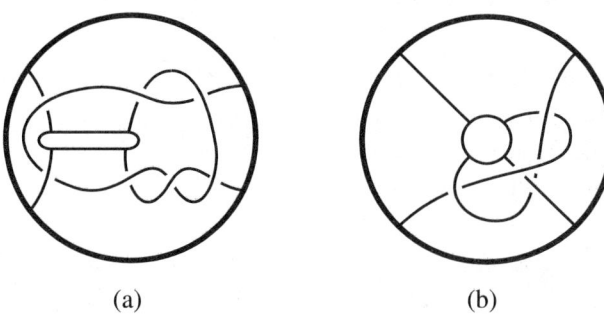

Figure 24.2

Denote by $L_i(n)$ the knot obtained from K_i'' by $1/n$ surgery on K_i'.

THEOREM 24.4. *A knot K in S^3 is hyperbolic and admits two toroidal surgeries $K(r_1), K(r_2)$ with $\Delta(r_1, r_2) \geq 4$ if and only if (K, r_1, r_2) is equivalent to one of the following, where n is an integer.*
  (1) $K = L_1(n)$, $n \neq 0, 1$; $r_1 = 0$, $r_2 = 4$.
  (2) $K = L_2(n)$, $n \neq 0, \pm 1$; $r_1 = 2 - 9n$, $r_2 = -2 - 9n$.
  (3) $K = L_3(n)$, $n \neq 0$; $r_1 = -9 - 25n$, $r_2 = -(13/2) - 25n$.
  (4) K is the Figure 8 knot; $r_1 = 4$, $r_2 = -4$.

PROOF. We first show that the knots in the list are hyperbolic and the slopes are toroidal. For L_1, $L_1(n)$ is a hyperbolic twist knot for $n \neq 0, 1$, on which the toroidal surgeries have been classified in [BW]. For L_3 this is due to Eudave-Muñoz, see [Eu, Theorem 2.1 and Prop. 2.2]. (One can check that $L_3(n)$ is the same as the Eudave-Muñoz knot $k(3, 1, -n, 0)$ in [Eu, Figure 25].) Since there is an isomorphism of S^3 which interchanges the two component of L_2, by Lemma 24.3(1) the only pair of toroidal slopes of distance at least 4 on K_2' are ± 2. Recall that $L_2(n)$ is obtained by $1/n$ surgery on K_2'. Since L_2 is amphicheiral, $L_2(n)$ is homeomorphic to $L_2(-n)$, hence if $L_2(n)$ is toroidal for some $|n| > 1$ then so is $L_2(-n)$ and K_2' would have a pair of non-integral toroidal slopes ($\pm 1/n$) of distance at least 4, which is a contradiction. Similarly it follows from [Wu2] and [GW3] that $L_2(n)$ is also nontrivial and anannular for $|n| > 1$, and hence is hyperbolic. Also, $L_2(n)(r_i)$ ($i = 1, 2$) is the double branched cover of the tangle obtained by gluing a $1/n$ tangle to the tangles in [GW1, Figure 7.4(d) and (f)], which is the union of $T(2, n)$ and $T(2, 3)$, and hence is toroidal when $|n| > 1$. This completes the proof of the sufficiency.

We now prove the necessity, so assume that K is hyperbolic knot admitting two toroidal surgeries $K(r_1), K(r_2)$ with $\Delta(r_1, r_2) \geq 4$. By [Go] the Figure 8 knot $L_1(-1)$ is the only hyperbolic knot in S^3 admitting two toroidal surgeries of distance at least 6, so we assume $\Delta = 4$ or 5. By Lemma 24.2 there is a homeomorphism $\varphi : (M_i(s), \{r_i', r_i''\}) \cong (E(K), \{r_1, r_2\})$ for some $i = 1, 2, 3$ and $s \subset T_1$. It is easy to see that $L_i(n)$ is either trivial or a torus knot when $n = 0$ or $(i, n) = (1, 1)$ or $(2, \pm 1)$. Therefore we only need to show that $s = 1/n$ because the slopes r_i can then be calculated using Lemma 24.1 and the Kirby calculus [Ro, p.267].

If Dehn filling on T_1 of ∂M_i along slope s produces a knot exterior $E(K) = M_i(s)$, then the meridian-longitude of K may be different from that of K_i'' on T_0.

We use (m'', l'') (resp. (m, l)) to denote a meridian-longitude pair of K_i'' (resp. K) in S^3.

CLAIM 1. *If $E(K) = M_1(s)$ for some s on T_1 of ∂M_1 then $s = 1/n$.*

Since the linking number between the two components of L_1 is 0, a p/q Dehn filling on T_1 produces a manifold $M_1(p/q)$ with $H_1(M_1(p/q), \mathbb{Z}) = \mathbb{Z} \oplus \mathbb{Z}_p$, hence $M_1(p/q)$ is a knot complement only if $|p| = 1$. It follows that $K = L_1(n)$, where $n = qp$.

CLAIM 2. *If $E(K) = M_2(s)$ for some s on T_1 of ∂M_2 then $s = 1/n$ for some n.*

As before, let $L_2 = K_2' \cup K_2''$. Let $M = E(K)$. Assume $s = p/q$ and $|p| > 1$. We have $K(m'') = L_2(s, m'') = K_2'(s) = L(p, q)$. Therefore by the Cyclic Surgery Theorem [CGLS], m'' is an integer slope with respect to (m, l), say $m'' = am + l$. By [GLu1] the toroidal slopes r_1, r_2 of K are integer or half integer slopes with respect to (m, l). Recall that $\varphi(r_2') = r_1$ and $\varphi(r_2'') = r_2$. Since m'' is an integer slope, r_1, r_2 cannot both be integer slopes, otherwise $4 = \Delta(r_1, r_2) \leq \Delta(r_1, m'') + \Delta(m'', r_2) = 2$, which is a contradiction. Also by [GWZ] they cannot both be half integer slopes.

Now assume r_1 is an integer slope and r_2 is a half integer slope with respect to (m, l). Since m'' is an integer slope, we may choose $l = m''$. Then $r_1 = p_1 m + l$ and $r_2 = p_2 m + 2l$, so $\Delta(r_1, m'') = \Delta(r_2, m'') = 1$ implies $p_1, p_2 = \pm 1$. But then $\Delta(r_1, r_2) = |2p_1 - p_2| \leq 3$, a contradiction.

CLAIM 3. *If $E(K) = M_3(s)$ then there is an integer n and a homeomorphism $\eta : (M_3(1/n), \{r_3', r_3''\}) \to (E(K), \{r_1, r_2\})$.*

By Lemma 24.3 $M_3(-7)$ is a solid torus, and the meridian slope m'' and the slope $r = -7$ are the only solid torus filling slopes on T_0. If $\varphi(m'') = m$ then $S^3 = K(m'') = K_3'(s)$ implies that $s = 1/n$ for some n, so $\eta = \varphi$ is the required map. If $\varphi(r) = m$, let ψ be the orientation preserving homeomorphism of M_3 given in Lemma 24.3, which maps m'' to r. By Lemma 24.3 ψ interchanges the slopes r_3', r_3''. Let $s' = \psi^{-1}(s)$. Then $\varphi \circ \psi : (M_3(s'), r_3'', r_3') \cong (E(K), r_1, r_2)$ maps m'' to m. As above this implies that $s' = 1/n$, hence $\eta = \varphi \circ \psi$ is the required map.

We now assume that $\varphi(m'') \neq m$ and $\varphi(r) \neq m$. Note that $K(m'')$ and $K(r)$ can be obtained from the solid tori $M_3(m'')$ and $M_3(r)$ by s filling on T_1, so they have cyclic π_1, hence by [CGLS] r, m'' are integer slopes of K. Choose $l = m''$. Since $\Delta(m'', r) = 1$, we may assume $r = 1/1$ up to rechoosing the orientation of l. The toroidal slope $r_3' = -9$ satisfies $\Delta(r_3', m'') = 1$ and $\Delta(r_3', r) = 2$, which implies $r_3' = -1/1$ or $1/3$ with respect to (m, l). The second is impossible by [GLu1]. Similarly the fact that $\Delta(r_3'', m'') = 2$ and $\Delta(r_3'', r) = 1$ implies that $r_3'' = 2$ with respect to (m, l). But then we have $5 = \Delta(r_3', r_3'') = \Delta(-1, 2) = 3$, a contradiction. □

COROLLARY 24.5. *A hyperbolic knot K in S^3 has at most four toroidal surgeries. If there are four, then they are consecutive integers.*

PROOF. By [GLu1, GLu2] a toroidal slope of K must be an integer or half integer, and if it is a half integer then K is a Eudave-Muñoz knot. By [T3, Corollary 1.2], if K is a Eudave-Muñoz knot then it has at most three toroidal slopes, hence the result is true if K has a half integer toroidal slope. Therefore we may assume that all toroidal slopes of K are integer slopes. The result follows if $\Delta(r, s) < 4$ for all pairs of toroidal slopes (r, s) of K. Therefore by Theorem 24.4 we need only show that if K is either $L_1(n)$ or $L_2(n)$ for some n then K has at most three integer toroidal slopes.

If K is the knot $L_1(n)$ in Theorem 24.4(1) then by [BW] it has exactly two toroidal slopes unless it is the Figure 8 knot, which has three toroidal slopes.

Now consider a knot $K = L_2(n)$ in Theorem 24.4(2) and let r be an integral toroidal slope of K other than r_1, r_2 in the Theorem. Since $\Delta(r_i, r) \leq 4$, r must be between r_1 and r_2. Denote by $M_2(p/q)$ the p/q filling on T_0 with respect to the preferred meridian-longitude pair of L_2. By the proof of Lemma 24.1(2), $M_2(-1)$ is the double branched cover of $Q_2(1)$. Using the tangle in [GW1, Figure 7.4(c)] one can check that $Q_2(1)$ is a Montesinos tangle $T(1/2, -2/5)$, therefore $M_2(-1)$ is a small Seifert fiber space with orbifold $D^2(2,5)$. Since $L_2(n)(-1-9n)$ is obtained from $M_2(-1)$ by Dehn filling on T_1 and contains no non-separating surface, it is atoroidal. Because of symmetry (L_2 is amphicheiral), $M_2(1)$ is homeomorphic to $M_2(-1)$, so $L_2(n)(1-9n)$ is also atoroidal. It follows that the only possible integer toroidal slopes of $L_2(n)$ are $j - 9n$ for $j = -2, 0, 2$. This completes the proof. (Actually it can be shown that $-9n$ is not a toroidal slope of $L_2(n)$ either, so it has at most two integer toroidal slopes.) □

The following corollary is an immediate consequence of Theorem 24.4.

COROLLARY 24.6. *Let K be a hyperbolic knot in S^3 which admits two toroidal surgeries along slopes r_1, r_2, and $\Delta = \Delta(r_1, r_2) \geq 4$. Then one of the r_i is an integer, and the other one is an integer if $\Delta \neq 5$, and a half integer if $\Delta = 5$.*

Although there are infinitely many hyperbolic 3-manifolds M with toroidal fillings $M(r), M(s)$ at distance 4 or 5, we have shown that they all come from finitely many cores $X(r, s)$ as defined in Section 21.

QUESTION 24.7. *Are there only finitely many cores $X(r, s)$ of toroidal Dehn fillings on hyperbolic 3-manifolds with $\Delta(r, s) = 3$? $\Delta(r, s) = 2$?*

We observe that the answer to Question 24.7 in the case $\Delta(r, s) = 1$ is almost certainly 'no'; here is an outline of an argument. Let N be a closed irreducible 3-manifold with a unique incompressible torus T up to isotopy. Let F be a once-punctured torus, regarded as a disk with two bands. It is intuitively clear that, for any positive integer n, by tangling the bands in a sufficiently complicated fashion we can construct an embedding F_n of F in N so that if $K_n = \partial F_n$, then $N - K_n$ is hyperbolic, and K_n cannot be isotoped to meet T in fewer than n points. Let $M_n = N - \text{Int} N(K_n)$, and let r, s on ∂M_n be the meridian of K_n and the longitudinal slope defined by F_n, respectively. Then $\Delta(r, s) = 1$, $M_n(r) = N$ is toroidal by definition, and $M_n(s)$ contains the non-separating torus $\hat{F}_n = F_n \cup D$, where D is a meridian disk of V_s. Hence, if we make sure that $M_n(s)$ does not contain a non-separating sphere, then $M_n(s)$ is also toroidal. Since the number of intersections of K_n with T is at least n, the triples (M_n, r, s) cannot all come from only finitely many cores.

Bibliography

[Be]   J. Berge, *The knots in $D^2 \times S^1$ which have nontrivial Dehn surgeries that yield $D^2 \times S^1$*, Topology Appl. **38** (1991), 1–19.

[BW]   M. Brittenham and Y-Q. Wu, *The classification of exceptional Dehn surgeries on 2-bridge knots*, Comm. Anal. Geom. **9** (2001), 97–113.

[CGLS] M. Culler, C. Gordon, J. Luecke and P. Shalen, *Dehn surgery on knots*, Annals Math. **125** (1987), 237–300.

[Eu]   M. Eudave-Muñoz, *Non-hyperbolic manifolds obtained by Dehn surgery on hyperbolic knots*, Geometric Topology (Athens, GA, 1993), AMS/IP Stud. Adv. Math. Vol. 2.1 (W.H. Kazez, editor), Amer. Math. Soc., Providence, RI, 1997, pp. 35–61.

[Ga1]   D. Gabai, *Surgery on knots in solid tori*, Topology **28** (1989), 1-6.

[Ga2]   ——, *1-bridge braids in solid tori*, Topology Appl. **37** (1990), 221-235.

[GT]   H. Goda and M. Teragaito, *On hyperbolic 3-manifolds realizing the maximal distance between toroidal Dehn fillings*, Alg. Geom. Topology **5** (2005), 463–507.

[Go]   C. Gordon, *Boundary slopes of punctured tori in 3-manifolds*, Trans. Amer. Math. Soc. **350** (1998), 1713–1790.

[GLi]   C. Gordon and R. Litherland, *Incompressible planar surfaces in 3-manifolds*, Topology Appl. **18** (1984), 121-144.

[GLu1] C. Gordon and J. Luecke, *Dehn surgeries on knots creating essential tori, I*, Comm. Anal. Geom. **3** (1995), 597–644.

[GLu2] ——, *Non-integral toroidal Dehn surgeries*, Comm. Anal. Geom. **12** (2004), 417–485.

[GW1]   C. Gordon and Y-Q. Wu, *Toroidal and annular Dehn fillings*, Proc. London Math. Soc. **78** (1999), 662-700.

[GW2]   ——, *Annular and boundary reducing Dehn surgery*, Topology **39** (2000), 531-548.

[GW3]   ——, *Annular Dehn fillings*, Comment. Math. Helv. **75** (2000), 430–456.

[GWZ]   C. Gordon, Y-Q. Wu and X. Zhang, *Non-integral toroidal surgery on hyperbolic knots in S^3*, Proc. Amer. Math. Soc. **128** (2000), 1869-1879.

[HM]   C. Hayashi and K. Motegi, *Only single twists on unknots can produce composite knots*, Trans. Amer. Math. Soc. **349** (1997), 4465–4479.

[L1]   S. Lee, *Exceptional Dehn fillings on hyperbolic 3-manifolds with at least two boundary components*, preprint.

[L2]   ——, *Dehn fillings yielding Klein bottles*, Int. Math. Res. Not. 2006, Art. ID 24253, 34pp.

[MaS]   D. Matignon and N. Sayari, *Klein slopes on hyperbolic 3-manifolds*, preprint.

[MeS]   W. Meeks III and P. Scott, *Finite group actions on 3-manifolds*, Invent. Math. **86** (1986), 287-346.

[Oh]   S. Oh, *Reducible and toroidal manifolds obtained by Dehn filling*, Topology Appl. **75** (1997), 93–104.

[Ro]   D. Rolfsen, *Knots and Links*, Publish or Perish Inc., Berkeley, CA, 1976.

[Sa]   M. Sakuma, *Homology of abelian coverings of links and spatial graphs*, Canadian J. Math. **47** (1995), 201-224.

[Sch]   M. Scharlemann, *Producing reducible 3-manifolds by surgery on a knot*, Topology **29** (1990), 481–500.

[Sct]   P. Scott, *The geometry of 3-manifolds*, Bull. London Math. Soc. **15** (1983), 401–487.

[T1]   M. Teragaito, *Distance between toroidal surgeries on hyperbolic knots in the 3-sphere*, Trans. Amer. Math. Soc. **358** (2006), 1051-1075.

[T2]   ——, *Toroidal Dehn fillings on large hyperbolic 3-manifolds*, Comm. Anal. Geom., **14** (2006), 565–601.

[T3]	——, *On hyperbolic knots realizing the maximal distance between toroidal surgeries*, J. Knot Theory Ramifications, **15** (2006), 101–119.
[Wu1]	Y-Q. Wu, *Dehn fillings producing reducible manifolds and toroidal manifolds*, Topology **37** (1998), 95–108.
[Wu2]	——, *Incompressibility of surfaces in surgered 3-manifolds*, Topology **31** (1992), 271–279.
[Wu3]	——, *Sutured manifold hierarchies, essential laminations, and Dehn surgery*, J. Diff. Geom., **48** (1998), 407–437.

Editorial Information

To be published in the *Memoirs*, a paper must be correct, new, nontrivial, and significant. Further, it must be well written and of interest to a substantial number of mathematicians. Piecemeal results, such as an inconclusive step toward an unproved major theorem or a minor variation on a known result, are in general not acceptable for publication.

Papers appearing in *Memoirs* are generally at least 80 and not more than 200 published pages in length. Papers less than 80 or more than 200 published pages require the approval of the Managing Editor of the Transactions/Memoirs Editorial Board.

As of March 31, 2008, the backlog for this journal was approximately 17 volumes. This estimate is the result of dividing the number of manuscripts for this journal in the Providence office that have not yet gone to the printer on the above date by the average number of monographs per volume over the previous twelve months, reduced by the number of volumes published in four months (the time necessary for preparing a volume for the printer). (There are 6 volumes per year, each usually containing at least 4 numbers.)

A Consent to Publish and Copyright Agreement is required before a paper will be published in the *Memoirs*. After a paper is accepted for publication, the Providence office will send a Consent to Publish and Copyright Agreement to all authors of the paper. By submitting a paper to the *Memoirs*, authors certify that the results have not been submitted to nor are they under consideration for publication by another journal, conference proceedings, or similar publication.

Information for Authors

Memoirs are printed from camera copy fully prepared by the author. This means that the finished book will look exactly like the copy submitted.

Initial submission. The AMS uses Centralized Manuscript Processing for initial submissions. Authors should submit a PDF file using the Initial Manuscript Submission form found at www.ams.org/cgi-bin/peertrack/submission.pl, or send one copy of the manuscript to the following address: Centralized Manuscript Processing, MEMOIRS OF THE AMS, 201 Charles Street, Providence, RI 02904-2294 USA. If a paper copy is being forwarded to the AMS, indicate that it is for it Memoirs and include the name of the corresponding author, contact information such as email address or mailing address, and the name of an appropriate Editor to review the paper (see the list of Editors below).

The paper must contain a *descriptive title* and an *abstract* that summarizes the article in language suitable for workers in the general field (algebra, analysis, etc.). The *descriptive title* should be short, but informative; useless or vague phrases such as "some remarks about" or "concerning" should be avoided. The *abstract* should be at least one complete sentence, and at most 300 words. Included with the footnotes to the paper should be the 2000 *Mathematics Subject Classification* representing the primary and secondary subjects of the article. The classifications are accessible from www.ams.org/msc/. The list of classifications is also available in print starting with the 1999 annual index of *Mathematical Reviews*. The Mathematics Subject Classification footnote may be followed by a list of *key words and phrases* describing the subject matter of the article and taken from it. Journal abbreviations used in bibliographies are listed in the latest *Mathematical Reviews* annual index. The series abbreviations are also accessible from www.ams.org/publications/. To help in preparing and verifying references, the AMS offers MR Lookup, a Reference Tool for Linking, at www.ams.org/mrlookup/.

Electronically prepared manuscripts. The AMS encourages electronically prepared manuscripts, with a strong preference for $\mathcal{AMS}$-LaTeX. To this end, the Society has prepared $\mathcal{AMS}$-LaTeX author packages for each AMS publication. Author packages include instructions for preparing electronic manuscripts, samples, and a style file that generates the particular design specifications of that publication series. Though $\mathcal{AMS}$-LaTeX is the highly preferred format of TeX, author packages are also available in $\mathcal{AMS}$-TeX.

Authors may retrieve an author package from the AMS website starting from www.ams.org/tex/ or via FTP to ftp.ams.org (login as anonymous, enter username as password, and type cd pub/author-info). The *AMS Author Handbook* and the *Instruction Manual* are available in PDF format following the author packages link from www.ams.org/tex/. The author package can also be obtained free of charge by sending email to tech-support@ams.org (Internet) or from the Publication Division, American Mathematical Society, 201 Charles St., Providence, RI 02904-2294, USA. When requesting an author package, please specify $\mathcal{AMS}$-LaTeX or $\mathcal{AMS}$-TeX and the publication in which your paper will appear. Please be sure to include your complete mailing address.

After acceptance. The final version of the electronic file should be sent to the Providence office (this includes any TeX source file, any graphics files, and the DVI or PostScript file) immediately after the paper has been accepted for publication.

Before sending the source file, be sure you have proofread your paper carefully. The files you send must be the EXACT files used to generate the proof copy that was accepted for publication. For all publications, authors are required to send a printed copy of their paper, which exactly matches the copy approved for publication, along with any graphics that will appear in the paper.

Accepted electronically prepared files can be submitted via the web at www.ams.org/submit-book-journal/, sent via FTP, or sent on CD-Rom or diskette to the Electronic Prepress Department, American Mathematical Society, 201 Charles Street, Providence, RI 02904-2294 USA. TeX source files, DVI files, and PostScript files can be transferred over the Internet by FTP to the Internet node ftp.ams.org (130.44.1.100). When sending a manuscript electronically via CD-Rom or diskette, please be sure to include a message identifying the paper as a Memoir.

Electronically prepared manuscripts can also be sent via email to pub-submit@ams.org (Internet). In order to send files via email, they must be encoded properly. (DVI files are binary and PostScript files tend to be very large.)

Electronic graphics. Comprehensive instructions on preparing graphics are available at www.ams.org/jourhtml/. A few of the major requirements are given here.

Submit files for graphics as EPS (Encapsulated PostScript) files. This includes graphics originated via a graphics application as well as scanned photographs or other computer-generated images. If this is not possible, TIFF files are acceptable as long as they can be opened in Adobe Photoshop or Illustrator. No matter what method was used to produce the graphic, it is necessary to provide a paper copy to the AMS.

Authors using graphics packages for the creation of electronic art should also avoid the use of any lines thinner than 0.5 points in width. Many graphics packages allow the user to specify a "hairline" for a very thin line. Hairlines often look acceptable when proofed on a typical laser printer. However, when produced on a high-resolution laser imagesetter, hairlines become nearly invisible and will be lost entirely in the final printing process.

Screens should be set to values between 15% and 85%. Screens which fall outside of this range are too light or too dark to print correctly. Variations of screens within a graphic should be no less than 10%.

Inquiries. Any inquiries concerning a paper that has been accepted for publication should be sent to memo-query@ams.org or directly to the Electronic Prepress Department, American Mathematical Society, 201 Charles St., Providence, RI 02904-2294 USA.

Editors

This journal is designed particularly for long research papers, normally at least 80 pages in length, and groups of cognate papers in pure and applied mathematics. Papers intended for publication in the *Memoirs* should be addressed to one of the following editors. The AMS uses Centralized Manuscript Processing for initial submissions to AMS journals. Authors should follow instructions listed on the Initial Submission page found at www.ams.org/memo/memosubmit.html.

Algebra to ALEXANDER KLESHCHEV, Department of Mathematics, University of Oregon, Eugene, OR 97403-1222; email: ams@noether.uoregon.edu

Algebraic geometry and its application to MINA TEICHER, Emmy Noether Research Institute for Mathematics, Bar-Ilan University, Ramat-Gan 52900, Israel; email: teicher@macs.biu.ac.il

Algebraic geometry to DAN ABRAMOVICH, Department of Mathematics, Brown University, Box 1917, Providence, RI 02912; email: amsedit@math.brown.edu

Algebraic number theory to V. KUMAR MURTY, Department of Mathematics, University of Toronto, 100 St. George Street, Toronto, ON M5S 1A1, Canada; email: murty@math.toronto.edu

Algebraic topology to ALEJANDRO ADEM, Department of Mathematics, University of British Columbia, Room 121, 1984 Mathematics Road, Vancouver, British Columbia, Canada V6T 1Z2; email: adem@math.ubc.ca

Combinatorics to JOHN R. STEMBRIDGE, Department of Mathematics, University of Michigan, Ann Arbor, Michigan 48109-1109; email: FRS@umich.edu

Complex analysis and harmonic analysis to ALEXANDER NAGEL, Department of Mathematics, University of Wisconsin, 480 Lincoln Drive, Madison, WI 53706-1313; email: nagel@math.wisc.edu

Differential geometry and global analysis to LISA C. JEFFREY, Department of Mathematics, University of Toronto, 100 St. George St., Toronto, ON Canada M5S 3G3; email: jeffrey@math.toronto.edu

Dynamical systems and ergodic theory and complex anaysis to YUNPING JIANG, Department of Mathematics, CUNY Queens College and Graduate Center, 65-30 Kissena Blvd., Flushing, NY 11367; email: Yunping.Jiang@qc.cuny.edu

Functional analysis and operator algebras to DIMITRI SHLYAKHTENKO, Department of Mathematics, University of California, Los Angeles, CA 90095; email: shlyakht@math.ucla.edu

Geometric analysis to WILLIAM P. MINICOZZI II, Department of Mathematics, Johns Hopkins University, 3400 N. Charles St., Baltimore, MD 21218; email: trans@math.jhu.edu

Geometric analysis to MARK FEIGHN, Math Department, Rutgers University, Newark, NJ 07102; email: feighn@andromeda.rutgers.edu

Harmonic analysis, representation theory, and Lie theory to ROBERT J. STANTON, Department of Mathematics, The Ohio State University, 231 West 18th Avenue, Columbus, OH 43210-1174; email: stanton@math.ohio-state.edu

Logic to STEFFEN LEMPP, Department of Mathematics, University of Wisconsin, 480 Lincoln Drive, Madison, Wisconsin 53706-1388; email: lempp@math.wisc.edu

Number theory to JONATHAN ROGAWSKI, Department of Mathematics, University of California, Los Angeles, CA 90095; email: jonr@math.ucla.edu

Partial differential equations to GUSTAVO PONCE, Department of Mathematics, South Hall, Room 6607, University of California, Santa Barbara, CA 93106; email: ponce@math.ucsb.edu

Partial differential equations and dynamical systems to PETER POLACIK, School of Mathematics, University of Minnesota, Minneapolis, MN 55455; email: polacik@math.umn.edu

Probability and statistics to RICHARD BASS, Department of Mathematics, University of Connecticut, Storrs, CT 06269-3009; email: bass@math.uconn.edu

Real analysis and partial differential equations to DANIEL TATARU, Department of Mathematics, University of California, Berkeley, Berkeley, CA 94720; email: tataru@math.berkeley.edu

All other communications to the editors should be addressed to the Managing Editor, ROBERT GURALNICK, Department of Mathematics, University of Southern California, Los Angeles, CA 90089-1113; email: guralnic@math.usc.edu.